THE
WOOD FIRE
HANDBOOK

THE
WOOD FIRE
HANDBOOK

The complete guide to a perfect fire
Vincent Thurkettle

SECOND EDITION

MITCHELL BEAZLEY

An Hachette UK Company
www.hachette.co.uk

First published in Great Britain
in 2012.

This revised and updated second
edition was first published in
Great Britain in 2019 by
Mitchell Beazley, an imprint of
Octopus Publishing Group Limited,
Carmelite House,
50 Victoria Embankment,
London EC4Y 0DZ
www.octopusbooks.co.uk
www.octopusbooksusa.com

Distributed in the US by
Hachette Book Group
1290 Avenue of the Americas
4th and 5th Floors
New York, NY 10020

Distributed in Canada by
Canadian Manda Group
664 Annette St.
Toronto, Ontario, Canada M6S 2C8

ISBN: 978-1-78472-619-5

A CIP catalogue record for this book
is available from the British Library.

Printed and bound in China.

10 9 8 7 6 5 4

For this second edition:
Art director Jonathan Christie
Senior editor Pollyanna Poulter
Production controller Emily Noto
Copy editor Helen Ridge
Proofreader Jane Birch
Illustrations Emily Maude,
    except for those on pages 74, 156
    and 171 by Vincent Thurkettle

Dedicated to my Mother and Father,
for allowing me a gloriously unfettered
childhood, with the freedom to roam the
wild woods from dawn 'til dusk, and make fires.

CONTENTS

INTRODUCTION 9

*

CHAPTER ONE   THE WOOD FIRES LIFESTYLE 13

CHAPTER TWO   THE TREES 39

CHAPTER THREE   WOODCUTTING 61

CHAPTER FOUR   BUYING FIREWOOD 77

CHAPTER FIVE   SEASONING LOGS 95

CHAPTER SIX   THE WOOD STORE 109

CHAPTER SEVEN   SPLITTING LOGS 131

CHAPTER EIGHT   THE WOOD STOVE & OPEN HEARTH 153

CHAPTER NINE   FLAME, SMOKE, EMBERS & ASH 177

CHAPTER TEN   CAMPFIRES & WOOD FIRE COOKING 197

*

APPENDICES 215
USEFUL CHARTS 216
INDEX 218
PICTURE CREDITS 222
ACKNOWLEDGEMENTS 223

INTRODUCTION

The genial cheerfulness of logs blazing on an open hearth has no equal; wood fibres hold summer's languid days to be spectacularly released in a winter fire's warm radiance. But while the aesthetic appeal of the flames' winding dance and the embers' glowing colours is certainly a driver for the growing popularity of wood fires, there are other reasons too.

Sitting in soft firelight, I have often wondered why I am so fascinated by fire. And I am not alone in this feeling either – people and cultures worldwide consider fire a strangely singular comfort. It can certainly be a useful and constant companion, but it is so much more. Fire warms, consoles, and touches the very hearts of us, like a loving, worldly wise relative. A fireside chair is a cosy nest in which to while away the time deep in conversation or lost in sleepy-eyed contemplation.

Fire warms, consoles, and touches the very hearts of us, like a loving, worldly wise relative.

I have spent a lifetime with wood fires. As a boy, I would roam the woods with friends building leafy dens, all of which were furnished with a small fire. At the age of 16, I left school and went to work in woodlands – I was at once at home in my new career. I loved working as the horseman's boy, chaining up large chestnut logs while the weathered horseman coaxed the massive horse to haul them out. I also helped a lovely old lady in the estate's Christmas tree plantation. She had once been a beauty queen and during World War II was put to work as a Land Army girl in the woodlands; she never left. Another of my duties was to light the woodsmen's lunchtime campfire.

I soon found myself managing forests, but ended up loosely tied to a desk. By way of escape, a new passion for hunting for gold emerged, which took me to some of the world's wildest places. The campfire skills I had developed as a boy became crucial: as in ancient times, my fire was central to life in these remote camps. Leaving my comfortable, well-paid office job, I opted to embrace a simple, wholesome life working with trees and hunting for gold – and spending much more time by my beloved fireside. Since I left work, I have never once wished I was back at my desk.

The idea of this book had been forming in my mind for many years. There are plenty of volumes and television programmes about making fires for wilderness survival, but very few that talk about how to make a good fire at home – for heat, or simply for the love of having a cosy log fire. I know that my general happiness is enhanced by spending time beside fires, and it concerned me that much of what I'd been taught by the old woodsmen seemed to be fading away from our collective memory.

We are so much the result of fire's influence that it is almost as if we evolved concurrently with it. Much of what makes humans different to other animals can be traced back to our relationship with fire. Long, long before people learned to make fire at will, our small, ape-like ancestors learned that fire brought many benefits. They learned that it was good to be near fire and cautiously formed a partnership that has lasted to this day; in that time fire has not changed or evolved, but we have.

Fire changed our physiology, as well as giving us the time to think, plan, and organize. In the wild, chimpanzees spend more than six hours a day eating their raw food and feeding times would have been similar for our pre-fire ancestors.

····································

Much of what makes humans different to other animals can be traced back to our relationship with fire.

····································

Cooking food begins to break it down so that digestion in the body is easier and the energy is absorbed more quickly and efficiently. As a result of the discovery and development of cooking, many modern humans consume their total daily intake of food in less than an hour. Our bodies changed as a response to the benefits of cooked food: we evolved to have smaller mouths and teeth, a shorter digestive tract, and larger brains. Fire freed up our days, releasing us from raw food consumption and, importantly, it also lengthened our days. Our early fire-using ancestors now had spare hours in which to nurture their nascent social development, and they could use a fire's light to push back each evening's unwelcome darkness. When fire was first tamed is not known; there is evidence of formal fire use from 790,000 years ago at a lakeside site in Israel, but what of smaller fires out on the open plain, where the wind and weather will have long since scattered any trace? The remains of my many boyhood fires have now melted back into the woodlands I played in and that was only 50 years ago, so what hope is there of finding the fire of a Homo erectus from almost 2 million years ago? Whatever the truth, fire has been with us for an unimaginably long time, helping to shape our bodies and minds; perhaps it is this fact, wired deep into our sub-conscious, that fuels our fascination with its flickering flames.

··

Healthy woodlands play a vital role in protecting archaeology and the increasingly important regulation of our water and soils. Additionally, in burning wood we may be using some waste wood for fuel that would otherwise go to landfill.

··

Good fires are the result of acquired knowledge and practice. In this book I hope to share all that I have learned through many years of trial and error and from the generous help of the fire-wise people I have met and worked with. I have hugely enjoyed my life with wood fires, both at home and in camps from the Australian bush to the Arctic sub-tundra. One would have thought that in the 21st century this most antiquated fuel would be facing obsolescence, but in fact the opposite seems to be true. The husbandry of wood fires is enjoying a renaissance as fossil fuel costs soar and people's desire for a little more self-sufficiency and sustainability grows. In our gadget-filled lives, we seem to have a yearning to relearn the simple living skills that were once commonplace – although fire was the television of our forbears!

Gathering and using wood for fuel, sustainably and in balance with the other benefits that woodland and forests bring, is a good thing for all of us. The fossilized carbon fuels stay locked deep underground, and we gently continue our fascinating two million-year relationship with the unique phenomenon of wood fire.

For many people, fire is celebratory, used for bonfires, barbecues, and candlelit dinners. For others, as in ancient times, it is an important part of home heating. More than ever before, we have a responsibility to use our wood wisely and keep our air clean. Happily, we can.

CHAPTER ONE

*

THE WOOD FIRES LIFESTYLE

*"We chopped wood for the night and carried it in;
dry beech sticks as brittle as candy ... Indoors,
our Mother was cooking pancakes,
her face aglow from the fire."*

—LAURIE LEE, *CIDER WITH ROSIE* (1959)

Adopting wood as a heating fuel within your home is in some ways like getting a new pet – wood fires are a lifestyle choice. Your fires will require housing, tending, feeding, and cleaning out. They will bring mess into your home and be something else to think about before you go out for the day, or go to bed.

But for all that, how many of us love our pets, take great comfort in them, and would not be without them for the world? I know with absolute certainty that wood fires have hugely enriched my life. The gathering and storing of wood, the bits of sawdust, leaves, and twigs that land on my carpet, the occasional cleaning out of the soft white ashes – I love it. I would do it ten times over to have the cheerful glow of a living wood fire in my sitting room.

However much I enjoy the flames and firelight, there is more to it than that – there is something else, something subtle and deep, that drives me. It is my love of everything involved, the total process: self-reliance, the people I meet, and being in touch with my local woodlands and the wider countryside. There is a feeling of touching nature herself – not as a tourist or a spectator, but through being actively involved, understanding our past, and being at one with this most ancient way of life – which brings an awareness and sense of "belonging" that is hard to describe. For me, this all forms part of the "wood fires lifestyle" and no other heating fuel comes close.

In this chapter I intend to look at the question of whether wood fires and their associated lifestyle are right for you: to consider what is involved and the wider benefits of being involved with wood as a heating fuel. Many people are, quite rightly, increasingly interested in environmental issues and there is a very strong environmental case for responsibly burning sustainably produced fuel wood. When burning wood in the modern world, I believe there is a moral, environmental, financial, and/or legal need to get the most heat from the least amount of fuel, and to learn good clean-burning techniques to keep smoke and particulate emissions to a minimum. When burning wood, there are two key things to know: that the heat energy produced per kilogram of seasoned logs is roughly the same for all tree species, and that dryness is everything. However, there is still so much more to know about wood fires.

ARE WOOD FIRES
RIGHT FOR YOU?

Before we look too closely at the wider benefits of a wood fires lifestyle, we need to first consider whether wood-fired heating is right for you. There are six fundamental factors to consider when choosing to burn wood:

— **Is wood as a fuel right for you?**
— **What level of commitment are you prepared to make?**
— **Do you have anywhere suitable to store the wood (and do you really want a pile of firewood there)?**
— **Is your home (and chimney) suitable for wood fires?**
— **Do you live in an area where wood burning in the home is permitted?**
— **Are you confident of being able to obtain the wood you need in your area (and at a price you are prepared to pay)?**

Just before considering each of these questions in more detail, I think it is worth noting that no heating fuel is trouble-free. Even the most modern and sophisticated heating systems are fallible; electricity disconnections and power cuts are common, many areas are without piped gas, and bottled gas comes with its own problems and inconveniences. Domestic heating oil is a little like wood in that you need to order it in advance, store a long-term supply on your property, and find room for the boiler that burns it. However, you still need electricity to make the oil-fired boiler work, so if there is a power cut you will have no heating. Coal has the advantage of being a concentrated source of heat energy, requiring a smaller storage area, but everything else about it is dirty or smells bad. The tang of coal smoke scent in your room is not pleasant, and the clinker (ash that has fused into a hard glassy slag) and ash must be cleaned out daily.

The point is that, although using wood fuel may present a number of inconveniences, none of the available alternatives is perfect either. I've yet to meet anyone who takes pleasure in a coal, gas, electricity, or oil heating "lifestyle" – these fuels are just commodities.

Is wood as a fuel right for you?

This first question really addresses the simple choice of "Can I be bothered?" You need to be very honest with yourself here. People often say "where there's a will, there's a way" and in wood heating this is certainly the case: whether or not your chimney is perfect, local logs are readily available, or there are suitable sheds in your garden, if you are really keen there will be a way to make it work. Your keenness and commitment are very important, but before making a decision you should mull over all the wider lifestyle benefits. If you already have a wood fire then of course you are a fair way down the decision-making process, but if you are currently thinking about buying your first wood-burning stove or opening up a fireplace, you should take the time to consider whether using wood fuel will become a joy or a nightmare. To help determine this, I would recommend talking to friends or relatives already subscribed to the wood fires lifestyle. Then, once you are certain that you want to embrace this antiquated and inconvenient heating fuel, you should aim to learn all you can as quickly as possible. After all, wood burning is not that difficult – until quite recently it's what everybody did.

What level of commitment can you make?

To a large extent, the more important wood is going to be in terms of heating your home, the more you have to think about. For many people it is enough to have the occasional log fire in an open fireplace to supplement the existing central heating. This may be to make the room cosy as a treat during bad weather, or to make it extra special at Christmas or for a dinner party. At this *basic level* of commitment you will always be able to find enough wood, storage shouldn't be a problem, and as long as you learn how to lay and light a fire everything should be fairly straightforward. I think that for some people a log basket loaded with "carefullest carelessness" is more important than the fire. The wood basket heaped with beautiful white-barked birch logs has an aesthetic charm all of its own and is a most appropriate decorative feature beside a summer's empty hearth.

A *mid-level* commitment would imply that wood forms a major part of the home heating plan, but is combined with other heating. This is the situation in my cottage. I have two electrical storage heaters that are set on low throughout the winter to help keep out the chill and damp, but I also rely hugely on an excellent wood-burning stove in my living room and an open fire in my front room. Using wood-fired heating in conjunction with another fuel option is a good choice for many people. You can fully enjoy a wood fires lifestyle, but do not have all your eggs in one basket.

Perhaps you are considering a *high level* of commitment – total wood-fired heating. If so, then you really do need to consider each of the six factors on page 15 very carefully and take professional advice on wood fuel supply and appropriate boiler systems. While total wood heating is not uncommon in rural North America and the heavily wooded countries of Europe, it is very uncommon in Britain.

· ·

I never try to keep my stove going all night as many people do; there are real problems and dangers with doing this, especially with regard to open fires, but also with stoves.

· ·

Do you have anywhere suitable to store the wood?

The importance of a well-designed wood shed and the fundamentals of good storage are covered in *Chapter 6*, but for now the question is: do you really have the room for all the wood fuel you need? You should assume that you will be buying "green" or "unseasoned" wood – recently felled and high in moisture – at least until you are sure of a trustworthy supplier of "seasoned" dry logs. So you are probably looking at having at least one year's fuel supply on your property, off the ground, and under cover. If wood is to be a significant heating fuel for you then this will probably mean a few tons. I'm focusing on wood as firewood logs here; if you are thinking of woodchip heating then this is in a different league and a purpose-built woodchip store is needed. If wood pellets or briquettes are your chosen fuel, then storage should be simple as they will usually be supplied in bags.

Is your home suited to a wood fire?

This may be the one question that you need professional help in answering. We will look at the hearth, wood-burning stove, and chimney later in this chapter, however if there is any doubt at all and it is not the current custom and practice to burn wood in your home, then taking a few hours of professional advice would be prudent. There is a range of people worth listening to; some probably won't charge for their time, others may. But if you are likely to be burning wood for many years to come, a little time and money invested at this stage could pay huge dividends later. Local professional people worth approaching could include a wood stove supplier, an architect, and a chimney sweep. Between them you could reasonably expect these people to give you a good overall view of whether your home and chimney are suitable for wood burning and whether it is reasonable to do so.

Is wood burning in the home permitted in your area?

This question is not as simple as it seems and you may well have already received the answer when taking professional advice about your home. Most clean air legislation is not focused on what you are actually burning in the home, but the emissions you are releasing into the atmosphere – especially the emissions of particulate matter. I have often heard people say "I would love to burn wood, but I can't because I live in a smokeless zone." If you have not actually checked what the legislation in your area requires, then this is a purely self-imposed restriction. Domestic wood-burning stove technology is now so good that there are stoves approved for use in smokeless zones and when burning properly seasoned wood correctly, the emissions from these approved stoves fall well within permitted levels. This is because once these stoves are hot and running normally they thoroughly burn all that is inflammable within the wood and no smoky emissions of soot particles or unburned hydrocarbons are released up the chimney. So in fact the question really is: "Am I limited in my choice of fire and stove by local emissions regulations?" Your local government environment department should be able to guide you on this.

Can you obtain the wood you need?

I have left this question until last, but in fact it may be even more important than your commitment to having wood fires that we considered initially. Nothing can be achieved if the supply of raw material is not available. I first had this piece of wisdom explained to me when I was a very junior forest manager and went to visit a canny old saw miller who worked in Rendlesham Forest on the east coast of England. His business was booming and he was clearly a most capable man. I have always been interested in success and took the opportunity to quiz him on what he believed was the most important factor in the continuing prosperity of his business. Almost without hesitation he replied "Supply!" He went on to explain that it didn't matter how good your equipment, staff, designs, marketing, and selling techniques were – if you didn't have supply, you had nothing. What was true for a booming sawmill business is also true for your home. Long before you start looking at beautiful wood stoves, or the more prosaic matter of whether your chimney needs to be lined, look very long and hard at the question of wood fuel availability in your area (*Chapter 4* looks more closely at buying firewood and how to find local merchants and suppliers).

THE WIDER BENEFITS OF A WOOD FIRES LIFESTYLE

For me this is where wood as a fuel, and in particular firewood logs, scores massively over all of its competitors. I have no doubt that a wood fires lifestyle brings many welcome wider benefits and should enrich your life through better health and fitness, and the development of new skills. There are no zero-environmental-impact heating fuels, and sustainably produced wood, burned efficiently, is as good as it gets. Possibly as important as the environmental benefits are the psychological ones – making new and interesting acquaintances, the sense of well-being felt after working among trees, or just sitting quietly in the soporific calm of firelight.

Fuel and furze day

The village in Norfolk where I live has many excellent qualities, one of which is that it has a small charitable trust managing an area of wooded common that was set aside over 200 years ago for the "poor of the village to gather fuel and furze". ("Furze" is *Ulex europaeus*, now more commonly called gorse, and was once a very important fuel; Thomas Hardy's rural poor had a whole industry devoted to gathering and supplying furze.) The trustees organize a working day from time to time to keep the woodlands and fence lines in good order, and the payment for the day's labour is a small load of firewood. Perhaps that doesn't sound like a good deal, but let me describe the last day I spent working with them and explain why I feel that it is. It was a Sunday morning in early January; the ground was hard with frost and a thick, damp fog hung heavy in the woodland. There were eight of us working that day and as each one drifted in with an axe or chainsaw, calling cheerily through the gloom and keen to get started, I was encouraged to get a campfire going with the dripping twigs, ready to boil a kettle and fry bacon. Determined not to lose face, I took great care to light the fire with one match, using precious scraps of birch and honeysuckle bark from the sodden woodland. The group worked on steadily, each at their own pace – we were not on piecework, there were no bosses or foreman to chase or scold us, and no challenging production targets to meet. And yet, by the time the kettle was whistling, a couple of tons of firewood had already been logged and split. We worked a relatively short day, but each of us was tired as we packed up and headed home, damp and dirty and already making plans for the next "fuel and furze" day.

People working together in friendship, filling a foggy woodland glade with life and cheerful bustle, meant a great deal to me. By being interested in firewood and happy to give up some time, my life was enriched in the following ways: some "free" wood, improved fitness and health, learning or maintaining forestry skills, an improved social network, supporting a local charity, supporting my village – and happiness. I enjoyed being out in the winter-scented air, the simple unhurried work at my own pace, swinging an axe, loading

the split logs into the back of a pickup truck – and I knew from the pleasing aches I had the following day that I'd toned almost every muscle in my body. The question of how it brought me happiness is harder to answer though, and touches on a subject that has become something of a holy grail in recent years.

Now I realize that this may all begin to sound somewhat removed from logs, but for just a little longer I want to focus on the life-enhancing quality that wood fires can bring – something so important to us and barely touched upon in more technical "wood-as-fuel" books.

Improving your "social capital"

Why, then, did a few hours of hard, unpaid labour make me happier? Humans are essentially social animals with a very deep-rooted need to belong, to be part of a group, to respect others, and, in turn, to be respected. We need to believe that what we are doing is worthwhile and that our future is something to look forward to. That day of steady firewood production – working, laughing, jibes, banter, the planning beforehand, and discussions afterwards – reinforced my place in the village community. We can further define this using the phrase "social capital", which describes an important concept. We can readily measure our material worth or "fiscal capital" – property, earnings, savings, shares, pension, etc – but how can we measure our social worth? The human networks we belong to, the interaction, value, and respect we share with family, friends, work colleagues, and neighbours all count towards our "social capital". For most of us there would seem to be a direct correlation between the breadth and number of social networks we engage with, our place and respect within each network, and our happiness. That "fuel and furze" day topped up my social capital account very nicely.

Apparently the UK's gross domestic product has gone up roughly tenfold since the 1950s and yet our level of happiness is, at best, about the same.

In the paragraphs above I've described a full wood fires lifestyle experience, but you do not have to be fully involved to reap the social rewards. Together with two neighbours, I once bought several tons of full-length beech logs. We had a really good time hauling the logs home and sawing them up, complete with an excellent picnic in the woods and some home-made cider (once the cutting was finished). From time to time I have to buy in ready-cut logs, and even then I sometimes meet the men who delivered them in the pub later. On walks with my children I would always have them on the alert for any kindling twigs or bits of driftwood we could take home – unusually practical and short-lived souvenirs of our day out. It all makes happy memories and builds some more welcome social capital.

Wood fires really should enrich your life, from the simple childlike pleasure gained by quietly watching your fire's mesmerizing dance while wind and rain lash the windows, to the memorable days out and friendships that can arise from a greater involvement. To be back in touch with the seasons and nature feels timeless – the only ancient skill remaining that is not just a hobby. You must be sure that wood heating is right for you, as discussed in the early part of this chapter – after that it is up to you to choose how involved you would like to be.

YOUR HEARTH, STOVE & CHIMNEY

It is best to approach the question of the type of fire you need as a simple appraisal. First of all, consider the status quo – whether the "do nothing" option is acceptable. In this case you are happy with the fireplace or stove you have and so no changes are required. Then again, you may decide that what you currently have is not quite right and you want greater efficiency or cleaner burning and need to improve your fireplace, stove, or chimney. The third possibility is that you are starting from scratch – your home is very modern and perhaps doesn't even have a chimney.

My first house was relatively modern and had been designed to burn coal. I rented this property and was not permitted to change the two open fireplaces, but as a forester there was no way I wanted to live with coal fires. I therefore had to accept the status quo in terms of the fireplaces and focused on changing the heating fuel. My answer to this conundrum was to burn small, very dry logs of only the best firewood – my woodpile was 90% beech, oak, hornbeam, and elm. This worked and kept my home warm enough. But on the very coldest days, and although it felt a little bit like cheating, I did perk up even these very good fires with a little sea coal. I lived 4.8km (3mi) from the coast and adored beachcombing after winter storms looking for amber and jet, but what I mostly found was washed-up coal, so I took the occasional sackful home.

Improving your fireplace

If you have an open fire but feel sure that things could be improved, then your new ideas and problem-solving list should include the following considerations. You should think about the possibility of raising the hearth; people with wood fires in old houses have found that this helps the fire to draw in air and, as the fire is a little higher, it is warmer to sit around. One man who took considerable care with this question believed the ideal height to be 25cm (10in) from the floor. If the fire smokes, fitting a hood in the fireplace may help. When coal became plentiful, in many old houses the large open wood-burning fireplace was closed up and fitted with a very small coal grate. If there is evidence of this in your home reopening the old fireplace is an option. I did this once in a Norfolk farmhouse. The small Victorian grate was pretty but useless, and the lintel above it showed clearly that the fireplace had once been much bigger. I knocked a hole through and peered in with a torch – it was glorious. I felt like a caver who had just discovered a group of prehistoric paintings, as behind the Victorian front was a vast cavern, the chimney bricks blackened from the wood fires of well over 100 years ago. My first thought was the possibility of hidden treasure, so I made the hole a little bigger and sent Daisy, my ten-year-old daughter, in with a torch. The ancient oak lintel had spells carved across it, so I assumed that further precautions against witchcraft might lie hidden in the chimney. Daisy was pleased to report no mummified cats, but sadly no hidden hoard either.

A massive improvement to consider with open fires is the possibility of creating an underfloor draft. This is a duct from the fireplace to the outside which feeds air directly into the fire.

This improvement hugely reduces the drafts across your room, as it stops the cold outside air from being sucked in through gaps around the windows and doors. The outside end of the ducting needs a grill to prevent unwelcome wildlife from also using it. Another major undertaking, but worth considering if you are doing any building work to the chimney, is fitting a back boiler to provide you with hot water from the fire. Finally, there is something wonderfully immediate about an open fire, with nothing between you and the flames, but perhaps the biggest efficiency improvement you can make is to fit an insert or a wood-burning stove.

....

I have seen some very good examples of an underfloor draft with a neatly polished, adjustable brass inlet grill either side of the fire. These need to be removable as curious children are apt to poke things into them.

....

The British seem to have been slow in adopting closed wood-burning stoves, preferring the blaze of a generous open fire, which is ironic as the open fire burns many more logs than a stove and we have for a long time been one of the least wooded nations in Europe. The wood fires enthusiast W. Robinson notes in his 1917 book *My Wood Fires and Their Story* that: "In many parts of France, Hungary, and Central Europe they have good wood fires, chiefly in closed stoves, which we might do well to imitate." Very much earlier the redoubtable Benjamin Franklin was of a similar view; desiring to improve upon inefficient household fires, in 1741 he set about designing an iron stove. While much modified, Franklin stoves are still widely available.

Whether you are looking at fitting a new wood-burning stove or simply replacing one that you are not happy with, give careful thought to the following advice. I strongly suggest that this is a time to buy the best quality you can afford; cheap and cheerful is not what you need with a wood stove. There is a wide range of good stoves available these days, from the almost futuristically modern to gorgeous, highly decorated cast-iron or ceramic traditional designs. The choice is yours – find something appealing that sits comfortably with the style of your home.

Some stoves have the option of being fitted with a water jacket and this may be something that appeals to you. As the stove will draw considerably less air than the open fire, your chimney will need to be fitted with a register plate. These plates seal off

the chimney while allowing access for the stovepipe. Make sure that the register plate fitted has an access hatch for an unlined chimney to be easily inspected and swept.

If in making your decision to embrace a wood fires lifestyle you decide to buy a wood boiler, then you most definitely need professional advice from a competent wood fires engineer. Modern boilers are superb and highly efficient, but in making your decision to use this technology to provide your heating, you must make a rigorous appraisal of your home's maximum and minimum heat load, the plumbing, the most suitable fuel type, and its storage. This is way beyond simply lighting a hearth or stove fire and you will need an expert's help to follow this option.

Starting from scratch

If you live in a modern home that was not built with a chimney and are really keen to have a fire, then there is still the possibility of having a wood stove. Fitting a wood pellet stove may be the easiest, but it may also be possible to have a log-burning stove with an insulated stovepipe running up the outside wall of the house. Again, I would strongly recommend that you take professional advice when doing this as strict building regulations govern the size, spacing, fitting, and height of the stovepipe.

It is hard to imagine nowadays, but not so very long ago only castles and manors had a chimney. In *The Cottage Homes of England* Stewart Dick states that: "It was probably not until the second half of the 16th century that chimneys in the smaller dwellings became common." Chimneys were something very special and most people, i.e. the rural poor, did not have one. In their rustic dwellings the smoke drifted and eddied until it finally found its way out – not necessarily a huge problem during summer, when the doors and shutters could be open, but possibly so in winter!

W. Robinson recalls staying in a farm cottage with no chimney: "The whole interior was black, and how people managed to live with it nobody knows, but they lived mostly outdoors." He was a great champion of wood fires and vigorously encouraged the use of wood over the "filth of Newcastle coal". This lack of chimneys was widespread throughout Europe. At the end of the 19th century the historian Alex Del Mar, while visiting the massive ancient Roman gold mines at Las Médulas in León, northern Spain, noted that: "In this house the hearth was composed of a raised platform of slate slabs, framed or bordered by logs. The fire was built in the centre of this platform, and over it was suspended an iron pot. Long wooden benches were arranged on each of the two sides of the hearth, and here the family sat while the pot boiled. The smoke, as in the other houses of this country, found its way out the best way it could. There was no chimney." We have of course learned that inhaling too much smoke may damage our health, but I'll give the last word to William Harrison, a rector in Essex, who complained in 1577 that the use of chimneys was making people soft: "Now have we manie chimnies, and yet our tenderlings complaine of rheumes, catarhs, and poses. Then we had none but reredoses, and our heads did never ake. For as the smoke in these daies was supposed to be a sufficient harding for the timber of the house, so it was reputed a far better medicine to keepe the goodman and his familie from the quacke."

Perhaps the fire's smoke fumigated these primitive homes, the people and their clothing, and this benefit offset the smoke's ill effects? But our "tenderlings" are constant and still complain! The likelihood is that you will have a chimney – so here are a few thoughts on it. If the chimney has been out of use for many years, then it needs a careful inspection before you light any fires. There may well be an old bird's nest blocking it, or bats living in it. You also need to check for loose or fallen bricks and that the chimney's lining is intact – no cracks to leak smoke into your rooms, or much worse cause a house fire. I would also have it thoroughly swept before use. If all seems well then consider whether it is best to have it lined before use with a wood fire and perhaps fit a cap on the chimney pot to prevent rain from entering the chimney. If this is not done, acidic sooty rainwater can corrode the register plate.

• •

Chimney sweeping is a profession and it is important to employ a good chimney sweep to ensure that not only the loose soot is removed, but also any tarry residues from the burning wood. Don't be tempted to try any "do it yourself" chimney-sweeping techniques. Pulling a bundle of gorse or thorn twigs up and down the chimney may be fun, but is probably ineffectual.

• •

Fireside paraphernalia
Clockwise from top left:
bellows; firedogs can
be purely functional or
ornate; stovetop fan; log
basket and gloves; cast-
iron fireback; fender.

How often your chimney should be swept will of course depend on how many fires you have and the quality of the wood you are burning, but in general having it swept professionally once a year is about right. Without wishing to be alarmist, the first sign that you are not sweeping the chimney enough could be the jet-engine roar of a chimney fire.

THE FIRESIDE BITS & PIECES

What can one say but that your friends and family need never struggle for Christmas gift ideas for you again! With the wood fires lifestyle comes a wonderful panoply of fireside accessories. A few are modern, but the majority are traditional and have been used around fires for centuries. When buying something new for your fire it is wise and prudent to be mindful of Sir Henry Royce's adage that "quality is remembered long after the price is forgotten". These items need to be well made; they are going to have a rough life. I've listed in the next paragraph those items that I've either seen or used, but once you start looking you may well find other pleasing knick-knacks or remnants of fireside antiquity.

Fireside tools and wood storage

Starting with the firewood itself, there are *log carriers* and *trugs* with which to bring the logs in from the woodshed and then *log baskets* and *log holders* to store them by the fireside. On the open fire the logs need some support to lift them slightly above their bed of ash and prevent them from rolling out; heavy iron supports called *andirons* or *fire dogs* are manufactured to do this. They are often quite beautiful and come in a very wide range of designs, from the gracefully patterned ornate to gundogs and mermaids.

A *fender* will certainly help to prevent logs and mess falling out from the hearth onto your carpet, but they also necessarily make a small barrier and the hearth seems slightly less accessible – I don't have one. You will need a *companion set of tongs, brush, shovel*, and *poker* to keep your hearth neat and in good order, and a pair of stout *heat-resistant gloves* to use these tools. I prefer the pokers that are designed with a hook, as these are good for rolling and positioning a log – wood fires don't need to be poked, only coal fires need poking to keep them bright. If you are uncomfortable with leaning in and putting your face and hair close to the fire when it needs some gentle blowing, then look out for *bellows* or one of the lovely and ingenious *blow pokers*.

Additional accessories

One of my favourite Christmas presents was a magnificent, cast-iron *fire back*. It was so heavy that I could hardly move it, but when it was in place my fire's flames illuminated a magnificent mediaeval hunting scene – the king's horse fidgeted in the light of the flickering flames. In the farmhouse I lived in before my current cottage, the fire back depicted Jesus and the Apostles at the Last Supper. If you choose to use long cook's matches or paper spills to light your fire, then you may like to invest in one of the purpose-made *match and spill holders*. You should also have a *metal bin* for taking out the ashes, but this can be stored away somewhere as you shouldn't need to use it very often. Another interesting invention is the *stovetop fan*. These are amazing pieces of engineering as the fan is driven purely by the heat differential between the stovetop and the air – no electrical leads or batteries are required. They move the warm rising air straight out from the fire and prevent a thermocline from developing in the room. Another device that many people find helpful in managing their stove, especially when cooking, is a purpose-made *wood stove thermometer*.

I keep a *heavy-bottomed kettle* set on *trivets* on my wood stove at all times. I enjoy having a little permanent free hot water for washing up or to top up the occasional bath. I am, perhaps unnecessarily, extremely wary of placing any pots or kettles directly on the stovetop in case they crack it. So, pots and my wood stove kettle stay on their trivets and heat up slowly. However, I also have a smoke-blackened *whistling kettle* over my open fire, which boils very quickly and makes excellent hot drinks. The old-fashioned way to do this was to have the kettle hanging from a *bender* mounted in the fireplace. The trivets

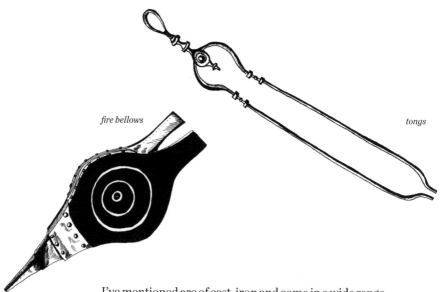

fire bellows   *tongs*

I've mentioned are of cast-iron and come in a wide range of delightful designs. Staying with cooking for a moment, I would recommend having a *toasting fork*, as English muffins and crumpets, or indeed simple sliced bread, toasted before a wood fire's embers are exquisite. Something once common, but seldom seen nowadays, is the *chestnut roaster* – a must for those who love the smell and taste of the autumnal sweet chestnut. I have also seen little *stovetop steamers* being used to prevent the room's atmosphere becoming too dry, or with a few drops of an essential oil to perfume the room. All being well, you will have avoided log types that are prone to sparking, but, just in case, it is wise to have a *spark guard* to protect your carpet and as a safety precaution for any time when you leave the fire. If you do not want to look at the empty fireplace during the summer, with its neatly raked bed of ashes ready for next winter, then a *fireplace screen* may interest you. I protect the Moroccan rug in front of my fire with a smaller, tatty old *fireside rug* – or sometimes a couple of sheepskins. And, still thinking of comfort, a *footstool* is great if you have a low hearth and it's not easy to warm your feet before the fire. My fireside is somewhat rustic and perhaps unusually I also keep a *small hand axe* and *saw* by the hearth for trimming up logs and cutting kindling sticks. Lastly, and perhaps the most obvious thing that is often overlooked, I have two *fire extinguishers* close to hand.

FURTHER BENEFITS
OF WOOD HEATING

Overall, wood heating is a good news story, although this is sometimes obscured in the melee of environmental politics, vested interests, lobbying, and opinion. I find that having an understanding of the environmental basics helps me to enjoy my life with wood fires. Mankind has enjoyed wood fires for as long as we have existed, so it is absolutely right that you should be comfortable with them. Having said that, there is also a modern responsibility to harvest firewood sustainably and burn it cleanly.

Our **wood fuel** source is trees, and we should understand their growth and management. The trees felled must be replanted in balance with the felling to maintain a truly sustainable system. A tree, woodland, or forest grows a certain amount each year and as long as we harvest no more than this accrued increment, then we have a system that can be maintained in perpetuity. The speed of growth varies between broadleaves and conifers, and between the individual tree species. A rule of thumb for most broadleaves is that normal temperate woodland is growing new timber at a rate of around 2–6m^3 (2.4–7ft^3) per hectare/year (0.8–2.4m^3/1–2.9ft^3 per acre/year). Some broadleaved species, such as eucalyptus, willows, and poplar grow faster. Conifers are in general much faster-growing too, pines and spruces being more in the region of 8–20m^3 (9.5–24ft^3) per hectare/ year (3.2–8m^3/3.8–9.5ft^3 per acre/year). In the northern boreal forests, conifer and broadleaf growth rates are slower. A cubic metre (35.3ft^3) of fresh wood weighs roughly 1 tonne. The planting of new areas of trees is particularly helpful as it widens the fuel source base. The aim of state forestry departments and professional forest managers is to achieve this state of equilibrium, or "normality", within existing woodlands and forests, and to plant new trees where appropriate.

The **carbon cycle** is important too, because, while wood is roughly 50% carbon, there is a very material difference between carbon in the world's active environmental cycle and that which is now fossilized. Trees absorb carbon while they are growing and release it again when they are burned or decompose.

When tree growth is in balance with the wood harvested, then we have an almost carbon-neutral system. I say almost carbon-neutral, as when harvesting wood fuel the equipment and fuel used to cut and haul the wood will in most cases have a fossil fuel footprint. However, when burning fossilized carbon fuels (oil, gas, coal) you similarly have an equipment and harvesting carbon footprint, but also the release of stored carbon that has not been in the active carbon cycle for millions of years. Growing trees also act as a temporary carbon store by locking it up (sequestrating) in their wood.

Firewood also tends to be ***bought locally*** and consumed close to source and has a proportionally small transportation footprint, making it overall a more sustainable material.

There are many wider environmental benefits associated with using wood as a fuel. History teaches us that people tend to care for things they value, and our forests and woodlands are more likely to survive and flourish if they are more useful to us than the land use alternatives, such as intensive agriculture or urban development.

History teaches us that people tend to care for things they value, and our forests and woodlands are more likely to survive and flourish if they are more useful to us than the land use alternatives.

If there is a viable market for firewood, then this helps with the costs of managing the sources, including street trees, wooded nature reserves, hedgerows, and all of the farm spinneys and copses scattered across the countryside that make the rural landscape so pleasing to so many.

The role of firewood in ***supporting the local economy*** is important too and often not given enough recognition. Many people in rural areas are fighting hard to maintain the jobs needed to keep their communities alive and viable. I know it's only one small step, but it is nevertheless a step in the right direction to buy your fuel locally. Economists say that the recirculation of money within a community is important. When buying most forms of heating fuel your money vanishes off into the far distance – sometimes even to another continent. If you buy locally, you are supporting your own community.

Let us take my own home as the basis for a very simple case study, as I feel it helps to illustrate many of the points I have made in the paragraphs above. I live in an old detached cottage with four small bedrooms and two main rooms downstairs. I burn around 5.4 tonnes of firewood each year on my wood stove or open fire. Around 5 tonnes of my wood is harvested from local woodlands – mostly within 1.6km (1 mile), and never more than 8km (5 miles) from my home. The other 0.4 tonne comprises scraps of wood I pick up throughout the year from arboricultural work and waste wood, most of which would otherwise be burned on a bonfire or go to landfill. The average growth rate of the broadleaved woodland near me is probably around 4 m³ per hectare/year (141ft³ per 2.4 acres/year). It therefore takes 1.25 hectares (3.0 acres) of existing and sustainably managed woodland to support my needs and, consequently, I am supporting the sustainable management of this woodland area. A few litres of fossil fuel are used each year to power the chainsaw that cuts my firewood and the vehicle that delivers it. I split my wood with an axe.

Staying with me as an on-going case study, I would note that I am currently building a new home. Its design is based on my old cottage, but with modern insulation and two wood-burning stoves – there is no open fire. I expect my new home to use less than half the firewood I currently use. I am really looking forward to being warmer while burning fewer logs more efficiently and cleanly than I am just now. For the record, I have also insulated myself better too. My grandfather wore a wool vest throughout the winter months and, recently, I followed his example and bought a Swedish-made Merino wool vest – it is amazingly good. I am now comfortable having my home 2°C (3.6°F) cooler, which saves even more firewood!

CHAPTER TWO

*

THE TREES

*"The best friend on earth of man is the tree.
When we use the tree respectfully and
economically, we have one of the greatest
resources of the earth."*

—FRANK LLOYD WRIGHT

When embracing a wood fires lifestyle, some knowledge of trees is good – it shows respect and helps to develop an understanding of the fuel that you have chosen to use, which is one of the greatest resources we have and one of the few that is truly sustainable.

Spending just a little time getting to know some different species of tree, where they are, and how they grow, will foster the desire to do the one thing that really matters – to leave at least the same amount of wooded resource as we have inherited, and hopefully more. It is likely that generations to come will need trees more than ever – not just for their shade and beauty, but as a vibrant reservoir of food, building materials, fuel, and clean air; we must have this earth partnership, this symbiosis. Of course we need the trees much more than they need us, but we can plant, nurture, and look after them and help to extend their range in both the rural and urban landscape. Humans care for that which they value, and we must reinforce our collective understanding of just how valuable our trees are.

LEARNING TO IDENTIFY TREES

This chapter aims to give you an overview of a range of tree species in relation to their usefulness and characteristics as firewood. I really enjoy being able to identify the different trees and knowing a little about each of them, but if this is not for you then I would suggest that you at least aim to learn enough to avoid the very poorest and poisonous firewood – after all, a log's dryness is the single most important thing about it when it comes to wood fires. But, even so, I would still urge you to spend a little time learning about your trees; in my experience (I taught tree identification at a forest management training centre for several years) most people quickly find it interesting as well as very rewarding. To get a grounding in tree identification you need a decent reference book and, if possible, some time spent walking in woodland with somebody who knows their trees. It is a practical subject and nothing beats standing in front of a tree and examining its features while they are being explained to you. If you don't have a knowledgeable friend or neighbour who can help, then consider a day out with a woodland club, society, or conservation group. Much as I love arboreta, I would suggest that they are not the best place to start learning tree identification. This may seem perverse as the trees there will probably be named, but the trees in an arboretum are usually a collection of the rare, colourful, and unusual – your firewood will come from your local woodlands and so that is where you should go to learn about the species you are most likely to encounter regularly.

The identification of logs

A woodsman will first try to identify a tree by its overall appearance. If there is any doubt, they will then inspect the summer leaves or winter buds, the nature of the twigs, and how the leaves or buds are arranged on the twigs. With many tree species the bark is distinctive. Pay particular attention to the bark when you are looking at a tree, as it becomes important later when you are trying to

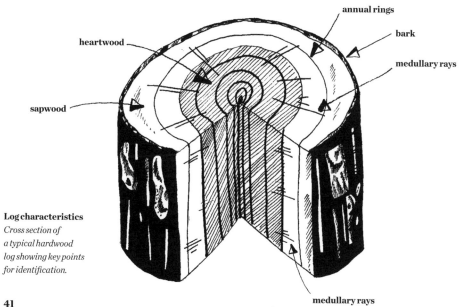

annual rings

bark

medullary rays

heartwood

sapwood

Log characteristics
Cross section of a typical hardwood log showing key points for identification.

medullary rays

41

identify your firewood. Logs are not easy to identify. After looking at any bark on the log you should then look at the annual growth rings and how clearly the inner heartwood and outer band of sapwood are divided; the medullary rays – ribbons of cells that radiate and transport the sap out from the centre of the tree – are sometimes distinctive. A log's colour, weight, and sometimes the smell of its bark or fresh-cut surface may also help identify it. The subject is huge – there are thought to be over 800 types of oak. Another problem even for experts is that many species frustrate the botanist's simple classifications: there are hard softwoods and soft hardwoods, there are deciduous conifers and evergreen broadleaves, some broadleaves have cones, and some conifers have fruits. But, thankfully, we don't need to know all of this to enjoy our wood fires. When you are confident that you can identify the wood of trees such as ash and oak, beech and maple, willow and poplar, and, critically, deciduous hardwood from coniferous softwood, then I would say you know enough to comfortably get by.

The scent of wood smoke

I have grouped the broadleaved trees in terms of their quality as firewood into excellent, good, and poor. Each wood species' other burning characteristics are noted – such as whether it throws sparks – and I have focused on a favourite subject of mine, the scent of the wood smoke. The conifers are looked at separately.

Not all wood smoke is equally pleasing and a few types are downright unpleasant. It's nice to see that Aristotle offers some guidance: "the kinds of wood, too, which contain more water, are less odorous than others," although he is not actually giving an opinion on whether the odour is pleasant or not. Of the dense woods, the ancient world seemed to particularly despise the smell of burning fig wood, which is odd as it smells vaguely of coconut. There are a few other woods I recommend keeping off the smoke list due either to their bad smell, such as dogwood and horse chestnut, or toxicity, such as laburnum, yew, rhododendron, sumac, euphorbia, and laurel.

Oak (*Quercus*)

Oak forms the mainstay of my fires. I split my logs small to hasten drying, as oak truly cannot be rushed, and seasoning for a year or more is best. The logs split well, although knotty ones will test your skill. The bark of most oaks is dark brown, rough, and fissured; the American red oaks have a much smoother bark. The wood has distinctive annual rings and a clear division between the pale sapwood and chestnut-brown heartwood. The medullary rays are large and clearly visible. I would once have laughed at anyone who told me that oak was good as kindling, but over the last few years I've started using it – when split into tiny sticks, this wood's great density gives real heart to the early fire. The wood is not inclined to throw sparks.

Just before discussing the scent of oak's smoke, I would ask that next time you split a piece of oak just pick it up and smell the freshly cut surface. Anyone watching may quietly decide that you have gone mad, but it has the most beautiful smell – sharp, yeasty, with notes of cut grass – I think it is my favourite of any cut wood. The smoke is strongly scented and it is easy to have too much in a room. I have also found this smoke to be one of the most aggressive and a face-full will quickly cause your eyes to sting and water. But get the balance just right and you will be rewarded with rounded, homely smells: mulled wine, cloves, citrus orange, and fresh brown bread.

Beech (*Fagus / Nothofagus*)

This is probably my favourite firewood. The logs split easily while the wood is green. The thin bark is smooth and a pale olive-brown colour. The annual rings are not clearly defined in this pale, fawn-yellow wood, nor are the sapwood and heartwood, but the medullary rays are evident as thin lines on the cross-cut surface, or flecking on the radially cut surface. Beech rarely throws sparks and makes good embers. I particularly like the rich, bright yellow flames you get on the open fire when burning beech. It is a traditional choice for wood-fired bread ovens and produces relatively little smoke.

Although beech is very good for smoking food, the smoke is gentle and well down my preferred list. It is easy on the nose and eyes and has a soft scent – perhaps newly mown hay and a slight hint of incense sticks.

Hornbeam (*Carpinus*)

This species makes such good firewood and is another favourite for the traditional bread oven – it would probably be great in a modern pizza oven too. This is a tough one to split as the wood is hard and commonly cross-grained; it is also slow to dry. The thin bark is smooth and olive brown. The wood is pale with no clear distinction in the annual rings or between the heartwood and sapwood. The greyish medullary rays are small. The overall look is of dull beech wood, but logs from the lower part of the tree may show evidence of fluting, which helps with identification. The wood rarely throws sparks and makes very good embers.

The smoke is largely a non-event. It is fairly sharp on the nose and eyes and generally seems to me to be neither sweet nor aromatic, although you may get a hint of caramel or burned sugar.

Elm (*Ulmus*)

It is an absolute puzzle to me why this wood is so ridiculed in poems pertaining to advise on the best woods to burn. It is a very tough wood, hard and cross-grained, and if you don't split it green you may have to get the chainsaw out! But I would have my woodshed full of it if I could find enough. My guess is that people were apt to burn it before it was properly seasoned. The bark is mid-brown and lightly fissured on younger trees, while on older trees these fissures develop and at first glance it can look similar to oak. The wood is dull brown when dried, but the sapwood is clearly distinguishable in freshly felled timber. Wych elm has greenish marks and streaks in the wood. The annual rings are clearly defined and for me the critical identification feature is that the xylem vessels (which carry water up the tree) are grouped into little wriggly strings on the cross-cut surface. The wood does not spark and the embers are excellent.

Elm has an interesting smoke. It is certainly very sharp on the nose and slightly eye-stinging, but, surprisingly, it is also fruity and fragrant, with suggestions of apple.

Field maple / hard maple (*Acer*)

I have included field maple (*Acer campestre*) in the excellent firewood category as the wood is particularly dense and burns really well – similar to the American favourite, variously called hard, rock, or sugar maple (*Acer saccharum*). Field maple often has a wavy grain and can be a difficult wood to split. The wood is a pale creamy white with no clear division between heartwood and sapwood, faint annual rings, and fine medullary rays. The field maple's bark is quite distinctive from other maples in having slight fissuring, which cracks into squares. The wood is a good, steady burn with good embers and is not prone to sparking. Annoyingly, I don't recall ever smelling its smoke.

Black locust (*Robinia*)

This tree is a native of North America but is now found across Europe, North Africa, and New Zealand. This hard, heavy wood is usually straight-grained and fairly easy to split; the wood is greenish when fresh-cut and later turns a golden brown. The bark is rough, deeply fissured, and stringy. The annual rings show and there is a clear division between the heartwood and sapwood. The wood rarely sparks and makes good embers.

I find the smoke to have a savoury, almost meaty scent. It is soft and pleasantly complex, with hints of wool or lanolin. When first smelled it seems to me slightly perfumed, something like sandalwood.

Hawthorn (*Crataegus*)

This is another of my favourite firewoods. While most commonly seen as a hedgerow shrub, it does grow tall with a single stem in woodlands. The wood is hard, but splits fairly well. The sapwood is creamy white and the heartwood is a delightful pinky-brown. Any smooth-cut surface at once seems polished. The wood is not prone to spark and burns bright.

The smoke is reputed to be good. The character Ratty, in Angela Huth's book *The Land Girls*, was lured by "the sweet smell of thorn smoke", and I have noticed this sweetness too on big outdoor bonfires, but indoors I have had trouble capturing it. I find the smoke earthy, sharp, nose-stinging and, sadly, rather nondescript.

Hickory (*Carya*) and Mesquite (*Prosopis*)

These American woods make excellent firewood: very tough, heavy, and difficult to split. Both give a high heat output and spark a little, but are particularly loved for their aromatic smoke – hickory is a favourite for smoking food.

Other excellent but less commonly available species

Yew (*Taxus*) is a hard, very dense wood and traditionally has been used as firewood. However, everything from the leaves and twigs to the wood and bark contains poisons (primarily the alkaloid called taxine and a cyanide-producing glycoside) and I have never been able to find out if these poisons are present in the smoke from burning yew wood. So, as a little smoke occasionally comes into my room from the fire and out of respect for my neighbours, I don't burn yew or other poisonous wood.

Box (*Buxus*) is extremely dense too and burns very well, but it always seems such a pity to burn this rare wood. I found the smoke of boxwood soft, sweet, with hints of bread, peanuts, perhaps even chocolate cake – which surprised me as the wet foliage smells unpleasant.

Laburnum (*Laburnum*) is another very dense wood that I have heard burns well but, as in the case of yew, every part of the plant is poisonous. Scientific opinion on whether or not burning toxic woods can cause harm is undecided, and so my recommendation would be to avoid them, as I do.

Rowan (*Sorbus*) is another very hard wood, but I have little experience in burning it – I may have been put off in the past by the fact that British folklore forbids burning it.

Dogwood (*Cornus*) is the favourite of some people, but as the smoke smells unpleasant, I avoid it.

Holm oak (*Quercus ilex*) is pretty special; it is often grown near the sea and I've found that some firewood may become available after a storm.

Bay (*Laurus*) makes delightful firewood. I was recently given some from a tree that had been killed by extreme winter cold. I could not resist seeing if the smoke was as pleasant as bay's leaves. It gave a delicious, rounded smoke that was easy on the eyes and smelled of spiced puddings and crushed peppercorns.

GOOD FIREWOOD

Alder (*Alnus*)

This often locally abundant tree is sometimes given a fairly poor rating as firewood. I'll be bold and simply say that this low rating is wrong. A friend of mine who lives by water meadows and burns alder most of the time makes wonderful fires with it. Writers often offer alder the faint praise that it makes good charcoal – well, it is much better than that. It is a fast-growing tree with a straight grain and is easily split, drying fairly quickly. When first felled the wood is pale, but the cut surface quickly darkens to a striking, rich red-orange colour – this dulls over time to brown. The sapwood and heartwood look similar. I don't remember noticing anything in particular about the smoke, and you will get the occasional spark.

Ash (*Fraxinus*)

So much praise is heaped on ash and I class it as a good firewood, but no more than that. It has the singular advantage of having a low moisture content when freshly felled, so was always a useful timber for people who had not prepared their firewood well in advance. I was pleased to read a very sensible piece of American research which ranked ash as the No. 1 firewood for people burning their wood green – as if that is ever a good thing – but it fell to No. 8 for people burning properly seasoned wood. I would say that the researchers got this exactly right. The wood is very pale, and while pinkish when freshly cut, it soon turns to light brown. The centre of the log sometimes has patches of dark brown wood. The annual rings are distinct, but the difference between the sapwood and heartwood is not. The bark is pale olive-grey and smooth, developing some fissures in older trees. The wood splits very easily and is not prone to spark; it produces very little smoke – perhaps another reason why people like it so much.

The smoke has a gentle, pleasing, almost sweet scent. Struggling to define it alone, I asked my now grown-up children to help with this one and they observed that it smelled of vegetable soup, perhaps vanilla, and hot pies.

Birch (*Betula*)

A particularly beautiful and hugely important firewood tree, especially in northern forests where conifers tend to dominate. The wood splits easily, other than those logs cut from near the stump where the tree tends to develop a wavy grain for strength. The bark is the most important identification feature and in mature trees it is white with large black diamond patches. The bark at the base of the tree is fissured into little dark chunks, often with pronounced fluting or "buttressing". The wood itself is white to light brown and finely textured; there is no clearly discernible difference between the sapwood and heartwood, nor are the annual rings clear. The split logs will dry fairly quickly and will spit a few sparks from time to time. Birch is very quick to rot if left anywhere damp with its bark still on. I particularly like birch as it offers more than most wood types: good clean firewood, masses of the special bark for fire lighting, a gorgeous scent when burned, and it just looks lovely in the woodshed or basket.

Birch smoke is probably the most aromatic of the common woods. The beautiful, tough bark is full of oil and you need a piece of wood with plenty of bark for room scenting, but care is needed. This oily bark is also highly inflammable and, apart from the fire danger, if you let birch bark burn with a flame in your room, you will soon find the air thick with floating particles of greasy soot. Blow out any flames and the smouldering log's deep blue smoke will give you a spicy, fragrant, fresh-cut straw scent.

Eucalyptus (*Eucalyptus*)

I've not had the chance to burn much of this wood, but I've liked what I have had. Split the logs as soon after felling as possible, as once dry the tough, cross-grained wood can be nearly impossible to split with an axe – some people just cut the logs very short and don't even bother trying to split them. The bark of most species starts smooth, but some soon exfoliate strips and chunks, giving them a shaggy appearance. The wood dries fairly quickly and I've found it to throw the occasional spark while burning.

The smoke is soft, not too eye-stinging, and has a pleasingly sweet fragrance with a predictable, but subtle, hint of eucalyptus.

Fruitwoods

The fruitwoods all make good firewood. The nature of the bark and timber varies significantly between the different tree types, but generally you will know what the wood is, having been told enthusiastically by the person you got it from. Fruitwoods are reasonably dense, fairly easy to split and burn, and throw few sparks, but their single most endearing feature is their smoke, so we will concentrate on that.

The famous "Woods to Burn" poem got it right when it said: "Apple logs will fill your room, with an incense-like perfume". The smoke of apple (*Malus*) is soft, sweet, and not eye-stinging. For me the scent brings to mind damson jam, cider, roasting chestnuts, lightly mulled wine, spice, and, interestingly, a hint of frankincense. My experience concurs with remarkable accuracy to the poem's "incense-like" description. I found the smoke of pear (*Pyrus*) surprisingly disappointing: thin and very faintly scented, perhaps a little like its fruit. However, cherry (*Prunus*) has a lovely smoke. I do not often get cherry wood so made a special effort to find some as I was keen to see if, as one poem notes, it really did scent my room of flowers. The result was not so much flowers as the absolutely gorgeous smell of an autumn kitchen, with the baking of scones and fruit pies. This is a very pleasant, soft smoke and, for me, the sweetest and fruitiest by some margin.

Hazel (*Corylus*)

Most hazel is grown as coppice wood and the logs are generally of small diameter. The bark is dark brown and the wood is very light brown. I find it just a middling firewood, but hazel does give a lovely, gentle smoke, offering homely smells: oat biscuits, hints of chocolate, and, not unreasonably, roasting nuts.

Holly (*Ilex*)

A strange firewood this one, said to burn while green and "waxy". The holly I have burned has been fine, but I wouldn't go out of my way to get any. It can be difficult to split, due to the many knots from the fine branches. The wood is greyish-white and the bark is olive green. If left in a damp pile the bark soon becomes very slimy and unpleasant to handle; this stickiness was once used to make "bird lime", a glue to catch small birds. The sapwood is not distinct from the heartwood, and the annual rings are not clear.

Holly has an aggressive, pungent smoke that rapidly stings the nose and eyes. In the smoke you may pick up hints of Olbas Oil, cough mixture, cough candy, and eucalyptus oil.

Plane (*Platanus*)

This tree has a particular tolerance to airborne pollution, which has led to it being widely planted in towns and cities; the source of any firewood may well be as a result of arboriculture work in parks or on street trees. The pale wood resembles beech, but has larger, reddish-brown medullary rays that show clearly. The sapwood is not distinct from the heartwood, but logs sometimes have a core of darker wood. On large trees the bark is very distinctive, with large dark grey or brown flakes falling off to leave pale yellow patches. This is not a firewood I have burned much, but it burned fairly well with a clear flame and did not spark.

Sweet chestnut (*Castanea*)

A fine tree, much loved for the shiny brown nuts in their spiny casings. The yellow-brown wood is at first glance easily mistaken for oak. In fact I would have put sweet chestnut in the excellent class, but for the fact that it is about 20% less dense than oak and is a terrible wood for throwing sparks. From time to time there is confusion when the word "chestnut" is used to describe firewood. Logs from the sweet chestnut and the horse chestnut are sometimes simply described as chestnut, and yet they are very different. Sweet chestnut logs, however, split well and are slow drying. The smooth bark on the young stem is at first purplish grey, later turning a silver grey. On older trees the bark is brown and deeply fissured. Sweet chestnut has a tendency to twist as it grows and a spiral grain may be evident in some logs. The pale sapwood is distinct from the heartwood and the annual rings are clear, while medullary rays are not. A good firewood for the wood stove, but not for the open fire.

Sycamore and other soft maples (*Acer*)

I'm grouping sycamore and the many other relatively fast-growing maples into one firewood category, as for the purpose of wood fires they are similar. Note that box elder is also included here, as it is actually a maple. The logs split fairly easily and dry reasonably quickly. The bark on most maples is smooth and light to dark grey, while older trees may have orange-brown fissures. The timber is yellowish white to light brown and sapwood is not distinct from heartwood; there may be small streaks of green staining. Annual rings are evident but faint, and the wood has thin medullary rays. The maples are not inclined to spark and make a good, general-purpose firewood.

Walnut / butternut (*Juglans*)

A good firewood, splitting reasonably well. It dries a little more slowly than other species in this good class. The bark on walnuts is generally smooth and grey, fissuring on older trees. The exception is the American black walnut, in which the bark is generally very dark brown to black. The wood is variable in colour, from greyish to rich dark brown, and marked with irregular dark streaks. The sapwood is very pale and clearly distinct from the heartwood; the annual rings are evident. Walnut is not inclined to spark.

I find the smoke very pleasant: fresh, tangy, and aromatic. It is sharp on the nose and eyes, but there are buttery, sweet hints of berries, fruit, and spice.

POOR FIREWOOD

I would only tend to use these firewood species in my wood-burning stove, as none of them is particularly good for an open fire. The first clue that you are holding one of these poorer firewoods is that the seasoned log will seem very light in weight.

Willow (*Salix*)

This tree is very widespread, over 300 species exist. It is very fast growing and will often be part of a mixed load of firewood. It dries very quickly and splits fairly easily, although because the wood is so soft an axe tends to get stuck in it, so it is better to use a splitting maul to break up willow logs. Most common willows have grey bark. The sapwood is white and heartwood is pinkish.

The smoke from willow smells vaguely similar to that of burning hay, but it is sharper, astringent, and more stinging on the eyes. I do like that it has a clean, fresh, vaguely herbal scent, suggesting tarragon, or perhaps marjoram.

Aspen / poplar / cottonwood (*Populus*)

While there are only about 30 species of poplar in the northern temperate regions, numerous hybrids and cultivars exist. Most are fast growing and become very large trees. Poplar's splitting, drying, and burning qualities are very similar to willow, and for me this is a firewood for the wood stove only. The bark is usually smooth and a dark green-grey, becoming deeply furrowed and cracked at the base as the tree grows. The wood of poplar is light in colour and may be white, greyish, pale brown, or even reddish. The sapwood is not usually distinct from the heartwood and the annual rings are not obvious. Poplar is not particularly inclined to spark, but as it will be mostly burned in the stove, it doesn't really matter. I have known many people who like to make their kindling from poplar.

Poplar's smoke is worth a mention – not because it is anything special, but as I am amazed that of all the different firewood types it has been singled out as "bitter" and generally dreadful in the traditional firewood poems. In my experience it is nothing of the sort and is really rather nondescript: sharp on the nose and soft on the eyes, definitely not bitter, but warm, soft, rounded, perhaps with a hint of sweetness, or ginger.

Horse chestnut (*Aesculus*)

A tree well known for its large spikes of springtime flowers and its autumnal conkers, the horse chestnut and its wood are nothing like the sweet chestnut. Horse chestnut splits easily and dries quickly. The bark is thin and dark grey to reddish brown; on older trees the lower bark breaks into large scales, which then fall away from the stem. The wood is creamy white or yellowish, and the sapwood is not distinct from the heartwood. The annual rings are not distinct. It burns fairly poorly, but in my experience does not spark. This is a firewood that I avoid because of its scent.

The smoke has a dull, vaguely unpleasant smell, which reminds me of old sacks rotting or smouldering.

Lime / linden / basswood (*Tilia*)

This tree is often seen planted in avenues and its subtle springtime flowers give nectar that bees use to produce a superbly flavoured honey. The wood splits well and dries fairly rapidly. The bark is usually dull grey and smooth on younger trees, soon developing fine fissures as the tree grows. The inner bark seems fibrous and stringy – I understand that it was once used to make a rudimentary rope. The wood is white or pale yellow, turning pale brown as it seasons. The sapwood is not distinct from the heartwood and the annual rings are not distinct either. It is this fine, uniform texture that has made it a favourite with woodcarvers. It burns fairly poorly but should not throw sparks.

Liking the honey, I was expecting the smoke to be good too – absolutely not! The smoke smells musty, of wet newspapers and old cooking oil. For me, the first whiff of this smoke also suggests seaweed, iodine, and salt – although in this I do get some more agreeable glimpses of Islay malt whisky.

Tulip tree / Yellow poplar (*Liriodendron*)

These trees are fast growing, often to quite a size; the namesake flowers are only prolific in hot summers. This wood splits fairly easily and the bark is grey with a shallow network of ridges. The sapwood is almost white and the heartwood a yellowish to pale olive brown. The annual rings are not distinct. The firewood is prone to sparking.

THE CONIFERS

This comprises the Gymnosperms, a large class of trees that includes pine, spruce, larch, fir, cedar, redwood, cypress, and hemlock: together colloquially known as "the conifers". The Angiosperms class includes all of the broadleaves.

Conifers are more of a challenge than broadleaves as firewood, but while most people would prefer to use firewood from broadleaved tree species, it is important to consider the conifers carefully. In many places conifers are the dominant trees and therefore the most freely available firewood.

The use of wood as a preferred heating fuel is increasing in some areas to a point where there are not enough hardwood logs to meet demand and people are now obliged to take a proportion of their firewood as conifer. And this is fine, as long as you understand that, in general, conifers are a subtly different wood fuel. When slow grown, such as in high or northern areas, the firewood is better than the faster-growing conifer trees in lowland sheltered valleys. As a huge generalization, they tend to split well, but are slow to dry and season thoroughly. They make good kindling, but as almost all conifers spit sparks they are best burned in a stove, and they mostly have a pleasant smoke scent.

The wood of conifers is structurally different from broadleaves in one particularly important way: it contains strong wood cells called tracheids, which have the dual functions of providing strength and conducting sap – these functions are differentiated in broadleaved timber's wood cells. Most conifer wood takes much longer to dry than similar hardwoods as the moisture flow with tracheids is not as free. Seasoning needs at least 18 months and many people opt for around three years before they consider the wood ready to burn. The sap itself is different too, being full of terpene compounds, and is much more resinous. What this means in practice for those people who are to burn a high proportion of coniferous softwood is that you will probably need much more storage space as these woods are not dense and need a long seasoning; also your chimney may need to be swept more often as conifer smoke is apt to produce an oily soot. A common complaint is simply that they burn too quickly. The answer to this is to burn the conifer logs in mixture with some hardwood logs. In fact, this is true when you have any poor-quality firewood to burn. Most reputable firewood suppliers will recognize these points and charge less for coniferous firewood.

Cedar (*Cedrus*)

These superb trees make the best of all the coniferous firewoods. The wood is quite dense, does not spit too much, and burns well; the smoke is particularly pleasant and richly scented. (Note that these are the true cedar trees seen in large parks and gardens, growing in the wild from the Mediterranean to the Himalayas.)

Pine (*Pinus*)

These trees make good conifer firewood and are locally common. They are slightly faster to dry than spruces and burn well, but are prone to sparking. The smoke is pleasant too, although very sooty.

Larch (*Larix*)

Another of the better conifer firewoods, but the propensity for larch to throw sparks explosively is well known. A fairly dense wood, it burns well and is again pleasantly scented. The larches are interesting in that they are deciduous; the spring flowers and newly flushed foliage are beautiful – as are the autumn colours. There are only ten closely related species of larch, but they are very widely spread. The three best known are European larch, Japanese larch, and tamarack, a larch species native to North America.

Cypress family and western red cedar (*Calocedrus, Chamaecyparis, Cupressocyparis, Cupressus,* and *Thuja*)

I have grouped this wide range of conifers as their firewood is similar. It is not dense and burns quickly, usually throwing plenty of sparks. In built-up areas this may be a frequent firewood type, as most commonly used hedging conifers are of these species. The scent of the western red cedar I find particularly delightful. I remember, as a 16-year-old boy, drilling large holes in log blocks of this wood for use as Christmas tree stands. There was plenty of friction from the drill and the workroom was soon filled with the most wonderful smell of mixed spice. Being hungry, this manifested itself with me at the time as the smell of my mother's bread pudding, which I adored. I still love this smoke today; it's all fruity spiced oranges, cloves, and rum.

Spruce (*Picea*)

The spruces make poor firewood, slow to dry and dull to burn. By the time they are fully seasoned there is little scent from them. I see these simply as a wood to burn for heat when there is little else – functional, but with no endearing qualities.

Fir (*Abies*)

I have never got on with fir firewood. Most firs are very fast growing, soft and pulpy with little resin pockets in the young bark. But these trees are widespread and provide plentiful firewood so may be useful to bulk up supplies. Some people are pleased to burn the Douglas fir and find it gives a relatively good heat output – I remember once struggling to get the wood to burn at all, but that was a long time ago and maybe I had not given it the very long seasoning time it needs. I do like the smell of firs, which I find rich, rounded, and suggesting orange marmalade.

Hemlock (*Tsuga*)

The first thing to mention is that these trees have nothing to do with the notorious "hemlock" plant used to poison Socrates and others – the only link is that it was once thought that the trees' foliage smelled similar. The wood logs are similar to spruce in the way they burn, so not a first-choice conifer firewood, but again may usefully fill an empty woodshed in lean times.

Redwood (*Sequoia, Sequoiadendron*)

These are perhaps the most beautiful and striking trees on the planet but make a pretty poor firewood. I have worked in plantations of the coast redwood and found the logs a mediocre burn – the bark of course is famously fireproof!

You may find other, rarer conifer species to try; juniper, for instance, is a joy to burn – a dense wood with a lovely scent. But, as I said earlier, the conifers in general are an important source of wood fuel and will be fine for the wood stove when well seasoned and, if possible, burned mixed in with other, better firewoods.

CHAPTER THREE

WOODCUTTING

*"Chop your own wood,
and it will warm you twice."*

—ANON

Woodlands seem to hold a special place in the hearts of many and, quite rightly, this leads to a desire to protect them from destruction.

This yearning to protect woodland will often manifest itself in the naïve belief that all tree felling is wrong; ironically, the movement "against felling" can *in extremis* lead to the decline, or total loss, of woodlands. As a forester I have often been asked: "If you like trees so much, why do you cut them down?"

WOODCUTTING
AIMS & HISTORY

The good woodsman should be neither timid nor reckless in managing woodland and making decisions to do, or not do, felling work. It is completely understandable that the woodsman's axe, or more likely today his chainsaw, is sometimes seen as a tool of destruction that in an ideal world should stay in the tool shed, but this is to miss the point of woodland management. Woodlands are robust and will survive periods of poor or indifferent management, but what they cannot survive is not being valued – there is always pressure on the land from agriculture or development. People care for the things they value, and if woodland is to regenerate contentedly through the seasons and survive generations of land-owning custodians, it must be useful to mankind and thereby have value.

Throughout history a woodland's usefulness was most commonly as a provider of food, fuel, and timber for buildings. As the beekeeper cares for a hive to enjoy a few jars of delicious honey, so the woodsman must care for the whole woodland to enjoy the crackling, cheerful warmth of a winter's log fire year after year. It is a wonderful thing that people feel protective about trees and woodlands, but it is important that this is not just benign sentimentality and that today's woodlands are seen as more than simply places of recreation or rural ornaments, breaking up a landscape of rectilinear fields and urban development. With a whole new generation of people taking an interest in wood as a fuel, woodlands are once again becoming increasingly appreciated for more than simply their aesthetic or amenity value. This, together with the careful and measured use of the woodsman's axe, should ensure that our woodlands continue to thrive.

This is a handbook of wood fires and I keep referring to wood as fuel, but woodlands are, of course, so much more than simple providers of fuel, timber, and food, and woodland management at the hands of experienced foresters and competent woodsmen recognizes this. A good woodland management plan will balance the needs of wildlife and game, the trees and other plant species, fungi, and the people using the timber, with other aspects such as the provision of shelter and screening, recreational activities, and the connectivity that the woodland provides within the wider landscape. In addition to this woodlands provide many less visible benefits for humans, which add further value and a greater need for ensuring that the forest ecosystem is functioning effectively; these include water and air purification, soil erosion regulation, archaeology, soil fertility, and carbon sequestration. In being sensitive to all interests and ensuring that many different groups of people care about and value the woodland it will remain in good health: vibrant, fecund, buzzing with life and energy, and growing plenty of wood – hopefully for centuries to come.

Human impact in woodlands

In seeking to understand woodlands it is important to have some perspective on the history of woodland management and the degree to which our woodlands and forests are natural. In fact, while there must once have been vast tracts of "wildwood", there are now only fragments of truly "natural" forest left – often in places inaccessible to people and their livestock, such as ravines. A large percentage of woodland has to some degree been influenced by the actions of mankind. However, this influence varies enormously, from the recent, man-made, fast-growing conifer plantations to the occasional tinkering with the ecological dynamics of semi-natural woodland and forest. There are perhaps three broad states of human activity that will affect the distribution of tree species and the internal structure of woodland. The first of these activity states is where there is no planned management and local people use the forest as a store and take a little produce when they need it, harvesting on an *ad hoc* basis from a seemingly endless supply. As long as the amount taken remains on a small scale and regrowth is encouraged, this can be a fairly sustainable system. However, in many developed countries this is largely a thing of the past and there is very little open-access, unmanaged woodland now remaining in which it would be reasonable or practical for people to routinely harvest to meet their day-to-day needs.

Grazing pressures

The second – much more noticeable – state is when the forest is used by people and also by their grazing livestock: where it is materially altered, but is still not managed. Woodland and forest will readily absorb the effects of light grazing, but prolonged overgrazing can be a disaster. This can be due either to differential browsing, i.e. the stock eat their favourite plants first, leading to the vegetation being dominated by plants that the livestock find unpalatable, or simply a general and widespread high pressure of browsing. The high-pressure browsing is a more serious situation as all the young tree and shrub seedlings are eaten, year after year, so that there are no young trees ready to take the place of those that are lost, whether through natural causes or harvesting. This insidious decline is most notable in regions such as sub-Saharan Africa and the lower Himalayan slopes, where in extreme cases actual desertification has resulted, but decline through over-grazing can happen anywhere.

I remember taking a group of students into a Scottish oak woodland and asking them to give me their opinion on the woodland's health. The woodland was beautiful, with evenly distributed, well-formed trees, but not one of the 18 young people with me noticed that there were no trees less than about 90 years old. The farmer had taken to sheltering his beef cattle in this woodland, as had his father and his father before him, and, in all that time not one tree seedling had gone unnoticed by the cattle. Nothing had been allowed to regenerate to replace the fine old oaks, as they became moribund and died. In fact, this lovely woodland was dying, but so slowly and so inadvertently that nobody had noticed.

And it is not just overgrazing by domesticated stock that can contribute to woodland decline. In recent years the population of the UK's wild deer – only two of which are native – has risen sharply. Woodlands are one of the main habitats that suffer, as the deer browse the field and shrub layer, stripping back the flora and tree regeneration, and thereby also reducing the populations of other animals that depend upon these plants, causing a general deterioration in the woodland ecosystem.

The third state of human involvement is when work is routinely undertaken to balance the negative effects of usage and ensure the long-term survival, health, and usefulness of the forest. The land managers will seek to maintain a tree structure or species balance that is beneficial to us and to biodiversity, through the planting and care of the preferred species, or the systematic removal of the less-favoured trees. Most woodland and forest across Europe and North America is now managed. Managed woodlands and forests often have improved access in the form of stone roads or grassy ride ways, areas of felling and subsequent replanting, and measures to control grazing animals and prevent the outbreak of uncontrolled fires.

Woodland management through the centuries

It must not be assumed that active woodland management is a modern phenomenon. It is clear from archaeological evidence that people have been managing woodland for a very long time, perhaps many thousands of years. The olive groves around the Minoan Palace at Knossos on Crete are a good example: little changed and still providing fuel and food as they have done for almost 4,000 years. We can be sure of this, as one of the wall frescoes discovered by archaeologist Sir Arthur Evans during his excavation of the site in the early 20th century clearly shows a grove of pollarded olive trees. But this site was lived on for at least 5,000 years before the Palace was built – so just how old are these enchanting, managed groves? Another glimpse of this antiquity is with the 5,800-year-old Sweet Track that crosses the marshy Somerset Levels in the south-west of England.

The nature of the poles and timber used in this track's wooden construction strongly suggests that some management had been taking place in the local woodlands, probably coppicing. When selecting the trees I cut for my winter firewood I like to reflect that, in taking this annual harvest, I am a part of a truly ancient tradition.

The management of woodland is highly desirable, but has never been easy due to the needs of people often changing faster than the trees can grow. In Britain this mismatch was particularly marked. Britain's recent policy history potted into one sentence goes something like: "Grow crooked trees for ship building – no stop, we need straight trees now; we need masses of underwood for hurdle fencing – no stop, we've just invented wire; firewood is vital as our only fuel – oh, no it isn't, we've found loads of coal; home-grown timber is vital in wartime … oh, … not any more, we have nuclear weapons."

In fact, after the move to iron ships, wire, and coal, and the end of trench warfare, woodlands were of so little value that by the early 20th century only 4% of Britain was wooded; thankfully, since then, tree planting has more than doubled this figure.

Notwithstanding the ongoing unforeseen fluctuations in our forestry policy, there are some basic, time-honoured principles observed in traditional management systems that still hold true. In practice these really boil down to four broad concepts:

— **Wood should not be harvested faster than the trees are actually growing.**
— **Young trees must be able to thrive in order to eventually replace old trees.**
— **Browsing by animals, wild or domestic, must be tightly controlled.**
— **Working practices must safeguard the woodland from the ravages of fire.**

Nowadays, this type of thinking is generically referred to as "sustainable" management. We want to believe that our woodland and forest will carry on in perpetuity and that, if possible, we will leave them for future generations in at least as good a state, if not better, than when we found them.

There is an interesting concept in professional and traditional woodland management that embodies these principles – it is called "normality". A woodland is said to have normality when all age classes are evenly represented and free to grow within its structure. In practice this is rarely achieved, but it is an ongoing goal of sound management to try and create, and then maintain, a situation in which the woodland has this perfect balance of infant, young, middle-aged, old, and veteran trees.

The management of woodlands and tree felling needs real knowledge and skill; it is not enough for someone to just "do their best". Well-meaning incompetence is not acceptable and may be dangerous. In the case of managing woodlands knowledge is essential as mistakes and blunders are sometimes not obvious for many years, or even a lifetime. Woodlands are the oil tankers of countryside management and changes in direction are best thought out carefully and made slowly. If you are contemplating any tree felling you should contact your local government office or a professional woodland consultant and take advice on the permissions or licence you may need.

SYSTEMS OF HARVESTING WOOD & TIMBER

A huge volume of excellent firewood comes from the management of individual trees in streets and parks, gardens, and hedgerows, but the majority of firewood is from our forests and woodlands. Wood produced through the management of these wooded areas can come from the removal of selected individual trees, as in thinning and continuous cover systems, or where all the trees are cut down in a selected area, as in clear-felling and coppicing. We'll look at the main woodland management systems, thinning, clear-felling, and coppicing, with a nod to pollarding as this was once common.

Thinning

Thinning describes the removal of poor quality or unwanted trees so that those remaining have the space to grow fully and mature. It is generally carried out at regular intervals throughout the life of the tree crop: roughly every ten years with broadleaves and every five years with conifers. I have watched gardeners thinning out trays of lettuce seedlings, and thinning trees is much the same thing – just on a larger and longer scale. The trees cut down during thinning operations are generally of poorer quality and smaller diameter than the trees left standing, and because of this they are usually sold into fairly low-quality markets; conifer thinnings largely go to paper and board mills, while thinnings from broadleaved woodland are often sold as firewood.

Clear-felling / clear-cutting

This is a management system in which all the trees in a given area are cut down at the same time, and is particularly common in conifer plantations. Tree seedlings are planted to form an even-aged block and then periodically thinned, throughout the rotation, to try and create a high-quality final crop. The entire stand of these economically mature trees is then felled. This management system is cheap and widely disliked due to the sudden apparent destruction of large chunks of forest; yet almost perversely there are many wildlife species, some quite rare, that enjoy and thrive in these large, temporary open areas. These clearings are soon replanted again with tree seedlings and this starts another crop rotation, ensuring that the overall system is sustainable at the forest scale. However, plantation forests managed on the clear-felling system, are often virtual monocultures. There are examples all over the world of monocultures planted and eventually failing due to the rampant development of some pest or disease. There is also evidence that the overall biodiversity of monocultures is poor and a growing understanding of the role of fungi in ensuring tree health – trees seem to grow better when in a mixed woodland. In the rush to plant vast areas of the fastest-growing trees, some foresters appear to have forgotten this lesson.

Coppicing

....

Coppicing is a long-established traditional woodland management technique. It is almost exclusively practised in deciduous woodland, as most broadleaves have the ability to regrow from the cut tree stump to form a coppice "stool". Only a very few conifers, for instance redwoods, will regrow from cut stumps. The very word "coppice" is delightfully flexible in the woodsman's lexicon, becoming a noun, verb, or adjective. There is a stand of coppice, (noun) and this indicates that the trees you are looking at have all grown from cut stumps. You can go out to coppice (verb) a woodland – that is to cut the trees down with the absolute, but unstated, indication that the cut area of woodland will be re-established by simply allowing the stumps to regrow, without any fresh tree planting. Then there are the coppice workers and coppice products, where the word is used as an adjective. Coppicing is, in reality, just small-scale clear-felling, yet to many people coppicing is the epitome of homely, traditional,

Coppice stages
Cut stump (fig. i); two-year coppice regrowth with many shoots at around 1.5m (5ft) high (fig ii); new tree stems at 15 years are ready to be cut as firewood at 10–15m (33–50ft) high (fig iii).

fig. i   *fig. ii*   *fig. iii*

sustainable woodland management, while clear-felling is suspiciously regarded as the preferred technique of short-term modern capitalism. The truth is that they are both perfectly reasonable techniques, as long as the felling done is in keeping with the landscape scale and type of woodland and the cut area is successfully restocked with trees – whether this is via planting or allowing the cut stumps to regrow.

In terms of firewood production, coppicing is an excellent technique. The felling is generally carried out in the winter and in spring the stump will produce several stems, which then grow rapidly as the tree's root system is already established. These stems grow into thick poles of a size that is easy to handle manually and cross-cut into logs. The coppice rotation (i.e. how long the trees are allowed to grow before they are felled) is much shorter than when growing forest trees, usually being something like 12–25 years, whereas a full-rotation tree timber crop is more likely to be around 65–150 years for broadleaves and 55–75 years for conifers. In terms of tree species, ash, hazel, hornbeam, oak, sweet chestnut, willow, and alder will all coppice really well; the maples are middling, while birch and beech make poor coppice. When managing coppiced woodland for firewood it helps to have a good density of stumps as this will lead to increased competition for light and taller, straighter pole growth.

The short cutting rotation followed by rapid tree re-growth, without the need of re-planting, is robust; the trees are unlikely to blow over and the cycle may carry on for hundreds of years. Coppicing, in fact, appears to rejuvenate many tree species. The one event that commonly spoils the young growth and wrecks the cycle is if the tender shoots are eaten off by domestic stock or wild animals. In fact, I would like to really stress this as I have seen more woodland wrecked by well-meaning coppice work than any other form of management. The problem is that the mammal browsing pressure on the stool regrowth is regularly underestimated and people believe that a flimsy fence or stick hedge will offer adequate protection. Hungry animals soon find their way in and decimate the delicate young shoots. Apparently Henry VIII decreed that felled areas must be well enclosed to prevent subsequent browsing damage – it's such a pity that we still haven't quite grasped how to do this properly.

Pollarding

A technique once much more common than it is today, pollarding has the singular advantage of allowing a coppice-type rotation, while also allowing livestock to freely feed and shelter within the woodland at all stages of the rotation. Many woodlands were once valued more for their ability to feed the owner's animals than for their timber, and so this was a great advantage. Pollarding is really just coppicing far enough up the tree's stem to prevent the young growth from being eaten by grazing animals. It is much more labour-intensive and dangerous than coppicing, as the woodsman is now doing the felling work several feet up in the air. Pollarding, if done correctly, also seems to prolong a tree's life and was often used to create huge old trees as ownership boundary markers.

FELLING TREES

The first thing I would like to be clear on here is that this is not a tree-felling handbook. However, I know that many people find tree felling interesting and want to know how it is done by those who do it for a living. If this is something that you are going to do, then I recommend most strongly that you attend a recognized tree-felling course and buy an appropriate chainsaw and safety clothing. Even if you intend to work using axes, billhooks, and bow-saws, you still need to be trained professionally before setting off to cut down trees. This is dangerous work and should only ever be approached with knowledge and respect. I could fill this book with cautionary tales from my own experiences – if you are going to fell trees, you must learn how to first.

Professional felling techniques

This is a brief summary of how a professional tree-feller will approach cutting down a tree. He will check his equipment and safety clothing, carry a basic first aid kit, know where his buddy is, and carry a phone or radio to call for help should an accident occur. On arriving at the tree to be felled, he will quickly clear the ground around

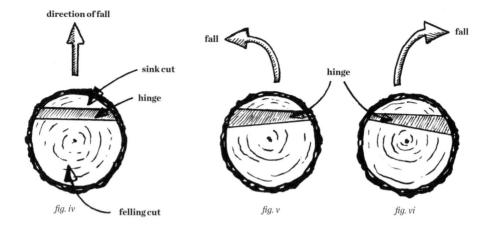

direction of fall

sink cut

hinge

fig. iv

felling cut

fall

hinge

fig. v

fall

fig. vi

Tree felling hinge
Cross section of cut stump with a sink cut (fig iv) and a hinge (fig. v and fig. vi) shows how the hinge is used to guide the direction a tree falls.

- -

Felling trees and the law
I believe that the management of woodlands and the regulations surrounding tree felling are now more closely controlled than they have ever been. It is no longer ethically acceptable, or legal, to cut large swathes of primal forest or wildwood with no regard to sustainability. Most countries even have strict rules for the harvesting of recently planted conifer plantations.

- -

the base of the tree, checking for stumps, old rabbit holes, or any other hazard that could cause him to trip or stumble while using the chainsaw. Next, he will examine the tree for any signs of rot, cracking, or embedded metal, and then look up to see if the tree is naturally leaning in one direction, or if its branches are tangled with the branches of another tree or overhead wires. The feller will look for any long dead branches in the canopy that could snap off and fall on him. He will then decide if the preferred felling direction is possible and, if not, where he intends the tree to fall. A competent feller will be aware of the weather forecast and, if the wind is likely to hinder the felling, will postpone the felling work.

The next step is to imagine exactly where he will be standing as the tree begins to fall and check that he has a clear emergency exit – at the very least, to be able to take a few steps away should the tree begin to fall badly. He would then make sure that he has close to hand any other tools that he may wish to use, say a hammer, wedges, or lifting bar. Once ready to start it is good practice to look again at the tree and ground, just to make sure you haven't missed anything. Some people like watching trees fall and the noise of a chainsaw attracts them, so every few minutes an experienced feller will look up and around to check that no one has wandered into the lethal felling zone to watch. People will sometimes do silly things, and when felling trees you must be constantly alert.

There are two basic cutting stages in felling a tree. The first is the "sink", "dip", or "mouth" cut (also called a "notch" or "face" cut in North America), and the second is the "felling" or "back" cut. On larger trees it may be necessary to "dress" the tree, that is to cut off any swelling or buttressing by major roots, before starting the felling process. The strip of retained wood between the dip and felling cut is called the "hinge", and this is hugely important. It is this hinge that guides the direction in which the tree will fall and prevents the weight of the tree from pinching down on your saw. When I am watching somebody felling with a chainsaw, I judge how skilful they are by how they have used their hinge. Anyone can saw a tree off and watch it fall, but making it fall where you want, time after time, is the real skill.

Processing a felled tree

Once the tree is felled, the work of breaking it down begins. There is a puerile woodsman's joke that we cut a tree down and then cut it up! Perhaps we spend too much time on our own. The felled tree is still not safe and the chainsaw operator will assess whether the main trunk is likely to roll as the branches are cut off, and if there is any significant tension or compression in the major branches. Once happy that he understands how best to approach the felled tree, the snedding (removal of side branches) and cross-cutting (cutting logs to length) work can begin.

Tree felling generally aims to cut the stumps as low as reasonably practicable so that they are not left as future obstacles throughout the woodland. However, when cutting coppice there are no general rules and each tree species should be approached differently. For instance, when cutting hazel the stump should be cut very low as this helps the young shoots to develop their own roots and reduces the amount of fungal decay within the stump. When cutting ash to coppice it is best to leave a high stump, as this will ensure plenty of strong shoot growth to form tall poles of this desirable firewood. Many believe the cut stump surface should slope so as to drain water away.

CHAPTER FOUR

*

BUYING FIREWOOD

"Logs to burn; logs to burn;
Logs to save the coal a turn.
Here's a word to make you wise
When you hear the wood-man's cries..."

—ANON, *PUNCH*, 1920

So starts a merry little poem that makes buying your firewood sound easy, but the truth is, when buying firewood there are no shortcuts. It is a matter of learning what you need, and then learning how to buy what you need. In this respect, it is surely the most complicated of all heating fuels but, ironically, this is what I like so much about it.

You can learn how to buy electricity, coal, oil, and gas in a few minutes, but learning how to buy your firewood is a deliciously fulfilling experience that will take much longer. In this chapter I will lay out what I have learned in a lifetime of wood burning, but I am only one man and would encourage you to ask the opinion of any experienced friends and neighbours who burn wood – local knowledge is always a wonderful thing. I would also suggest that how much time and thought you give to your wood buying should be directly proportional to the volume of wood you expect to burn. If wood is to be the primary heating fuel for your home, then the buying deserves a huge amount of very careful consideration. If, however, you are only buying a few logs for the occasional open fire then the buying requires little more thought than when buying a few bags of barbecue charcoal.

Most people selling firewood are trying to build and maintain a good base of recurring customers for their business – but not all. There will always be a scattering of fly-by-nights looking to cash in on the unwary. In this commodity, the maxim *caveat emptor* ("let the buyer beware") is particularly appropriate. Several years ago I dropped in on a friend one frosty Sunday morning. She had a lovely old cottage, but as I entered her home I could see that all was not well. Her rooms were full of smoke and she looked dishevelled, with black-smudged hands, and was clearly tired and exasperated. She was relieved to see me and asked if I would kindly light her fire. She said that she had been trying to light the stove for hours, but was now giving up as she was useless at fires and was going to get the stove taken out. I stooped to pick up one of the logs that she was trying to light – it was oak, sodden with sap and probably from a tree felled within the last month. I told her that these logs were non-inflammable and that you couldn't set fire to them with napalm. The wretched merchant had told her that the wood was ready to burn.

FACTORS TO CONSIDER
WHEN BUYING FIREWOOD

How to approach the issue of buying firewood comes down to a few basic questions surrounding the type of wood fuel, availability, degree of processing, quantity needed, and storage facilities. One of the things most people like about wood fires is the earthy simplicity; another is the rare sense of being at one with our ancestors, but the way we buy firewood today is not always the same. Traditionally, firewood was purchased by volume rather than weight – much more sensible. Thankfully, many merchants are moving back to selling by volume. This is particularly important when buying properly seasoned or kiln-dried logs, where the low moisture content makes the logs much lighter than "green" logs. If you don't have the storage room to season your logs, then buy them pre-dried. Take any stress out of this purchase by learning what you actually need and who can provide it at a fair price. Once you have developed a reasonable familiarity of your own, then log buying can become another pleasurable part of the wood fires lifestyle.

What type of fuel do you need?

This is the first of the big buying questions and the answer is largely dependent upon the type of fire you have. Here we are only looking at firewood as logs. We will look at the other forms of wood fuel (briquettes, pellets, and woodchips) later in this chapter.

For open fires
If you are buying wood to be burned on an open hearth then probably the size of log is not critical, but you would be wise to try and avoid tree species known to spit and emit sparks, such as larch, pine, and sweet chestnut. Also, it will be very hard to get a good open fire going solely with the less dense, fast-burning species such as poplar, willow, and possibly birch. But note that the density of a wood is to some extent dependent on where it has grown. I have sat around really excellent open fires in Canada and Scandinavia that were burning very slow-grown

birch, spruce, and pine. Once you know your supplier and learn which species you prefer, you can begin to ask for your favourites (see *Chapter 2* for guidance on my favourite firewood species). In practice, you will rarely get a delivery of all the woods you want and so will learn to use your first and second choice species in mixture. I mentioned log size earlier and, while this will depend upon the size of your hearth, there is much more flexibility with an open fire than when you have to get the logs into a stove's firebox. You may be lucky enough to have a large old fire-grate, perhaps even an inglenook fireplace; if so, you should be able to enjoy experimenting with a wide range of log sizes.

For wood-burning stoves and boilers
Burning wood within the firebox of an enclosed stove or boiler is easier than on the open fire. Sparks are not an issue and there is less skill needed to mix the dense and less dense woods and to achieve a pleasing balance of flames and embers. Even so, you still have to light and maintain the stove fire, and a range of log lengths and diameters will help you to do this. I start by measuring the longest and fattest log I can reasonably get into my wood stove and then make sure I have a full range of sizes below this maximum – they all have a role in starting and tending a perfect fire. For my wood stove I use a very wide range of tree species; even the utility poplar and willow will burn fine, and the smoke from smelly horse chestnut or dogwood never enters the room. When buying for the stove, size and dryness are important but you can be a bit more relaxed about the tree species.

What firewood is available?

It is important to be realistic. On his wall my father has the sagacious little saying: "Happiness is wanting what you have, not having what you want." This is especially important when buying firewood. There is little point in getting all excited and deciding that your fires will burn a crème de la crème mixture of fruitwoods and hornbeam if there isn't any growing near where you live. It is a good thing to know what you want, but "availability" is about having a reasonable expectation of what logs you can get, not simply accepting what the first merchant offers you.

The firewood market is maturing and there are now fully professional companies offering consistently good-quality logs. These people will tell you the volume and average moisture content of your logs accurately – they want you to be a repeat customer – and work is ongoing to develop industry quality standards. An awful lot of firewood is sold on the "grey" market, and that's fine too, especially if you are buying logs "green". It just takes more judgement to decide if the rough load delivered is fair for the price and you need some knowledge of tree species. Some of my favourite days have been going out into the forest with my children and a picnic in order to collect our own firewood. The children would ride in the trailer as soon as we were on the forest roads and then they'd bump along happily until we reached the heap of wood that I'd bought. I would saw the trunks into short poles and then the children would race to load the trailer, urged on by the promise of some chocolate on the way home.

How much work are you willing or able to do?

Always determine your own level of involvement by what you feel you will enjoy doing and what your circumstances will allow you to do – don't be swayed by the belief that cutting your own firewood will save you lots of money. It may seem cost-effective on paper, but you will work hard for the money you save. It's also important not to feel any pressure to suddenly be a woodsman. I've never heard of anybody who burns coal suddenly deciding they want to learn how to mine it at weekends. If what you want is a lovely wood fire burning sustainably produced logs, who actually processed the wood is of no consequence to the fire. Involvement is a purely personal decision, from having all of your logs delivered ready-to-burn, to doing everything yourself. There is a lot of hard work in turning a tree's stem and branch wood into split and seasoned firewood logs, and then further work in delivering them. Steady exercise is good for us and many welcome the chance to have input in producing one's own firewood. If, like me, you enjoy every stage of producing firewood, it's time to get trained, properly insured, and to the woods to cut your own.

How much firewood do you need?

As in other aspects of wood burning, this is a judgement and will vary depending on personal preferences and how severe the winter is. Your stove supplier should be able to give you some initial guidance on how much wood you may need in the average year. From this you will be able to calculate the size of wood store you need. If the store is to be used for seasoning next year's wood as well, it will need to be at least double the size.

To use my cottage again as an example, in order to fuel my open fire and wood-burning stove occasionally in the summer and on chilly spring and autumn evenings, as well as keeping the stove alight all day long in the 4–5 months of winter, I use about 9m³ (318ft³) (stacked volume including air space) of firewood a year. My stacking is fairly rough, as I want plenty of airflow through the woodpile, and I gather my kindling as dry twigs throughout the year.

So, a rule of thumb would be something like: aim to store 3–4m³ (106–141ft³) if you are going to have occasional fires throughout the year, or 8–12m³ (283–424ft³) if you expect your fire to be lit on most days through the winter as well as the chilly ones in spring and autumn. This assumes a normal mixture of hardwoods, as the log species is also important here. A cubic metre (35.3ft³) of something like oak or beech will give out around 50% more heat than a cubic metre of alder or willow.

How much wood storage room do you have?

It is essential to have enough storage room, but even better to have too much. Once the instinct to squirrel away as much wood as possible really bites, it is easy to get carried away and forget that it all has to be stored somewhere. Wood storage is discussed in *Chapter 6*, but for now let's just note that over-buying often leads to at least some of the logs being stored badly somewhere temporary. Judging how much wood your store can take should only really be an issue in the first year. After that you will know what the store will hold and so how much to buy in. As firewood often comes in awkward shapes, with variable moisture content, rules of thumb are difficult, but if all of this is new to you the following should be helpful in calculating how much wood storage room you have. A solid cubic metre ($35.3ft^3$) of freshly felled wood will weigh very roughly 1 tonne. The air space within a stack of cylinders is 9–22% of the solid volume. The rough and randomly shaped firewood will not stack anywhere near as neatly as cylinders and so there will be a greater air space within the heap – generally something like 35–45%. A further subtlety is that fresh wood will shrink by up to 20% during drying. Knowing this should prevent you from being puzzled when you notice that your once-full wood store develops a large gap at the top.

Is it best to buy by weight or volume?

I mentioned earlier that I thought that people in the past had a much more simple and informed approach to buying firewood, and for a time it seemed we had lost our way a little, as most buying was done by the "ton/tonne". The wood-buying knowledge our ancestors had was largely forgotten at the time when most households moved away from wood as their heating fuel to coal. Coal is fairly uniform and is generally bought by weight. But part of the fun of burning wood is that it is not an exact science.

Buying firewood **by weight** is absolutely fine when you have chosen to buy wet, green wood. Also, if you want to buy a large quantity, say a lorry-load, then this is the easiest way of buying the wood as the lorry can go over a weighbridge and the driver then gives you the weight ticket. As long as you were expecting the wood to be wet, the price reflects this, and you have the time and storage area to season the logs, then in many ways buying your firewood by weight is the easiest and most straightforward method as there are no concerns over moisture content or how tightly the logs have been stacked.

Buying firewood **by volume** is definitely becoming more common again. Merchants selling logs that have been dried and are ready for burning – whether properly air-seasoned or kiln dried – will usually opt for a volume sale. Otherwise, and very unfairly, their wood looks expensive if sold by weight. If you do want your logs to be ready-to-burn, then you can do so with more confidence if buying by a specified volume measure – a cubic metre ($35.3ft^3$), a load, a large sack, or crateful. Note

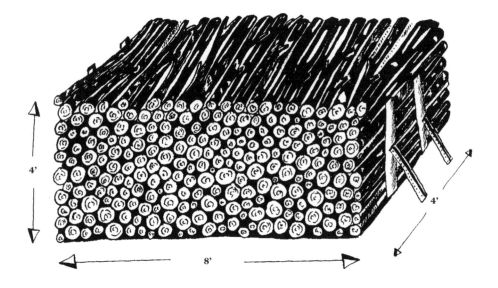

Cord of firewood

A standard cord of firewood has a total volume of 3.6m³ (128ft³).

whether the logs are stacked neatly or thrown in loose – it makes a big difference to the solid volume you are getting. Also, ask what species you can expect – a cubic metre (35.3ft³) of the denser hardwoods contains far more heat energy than the same volume of most conifers or species like willow, poplar, or alder. In the firewood vernacular a "good" load is one that the seller feels is slightly generous. This may be that the logs are packed more tightly than usual or that the top of the load is domed, thereby giving you a little more wood than the strict cubic measurement would allow.

The traditional unit of volume measurement for firewood is the "cord", which is still used in some parts of Britain and widely in North America. A "standard cord" is a heap 2.4m (8ft) long of 1.2m (4ft) firewood billets, stacked 4ft high. As a boy I was told that the logs should be packed so tightly that a squirrel could not get between them. The great advantage of this system is that everybody knows exactly what they are getting – 3.6m³ (128ft³) – no fussing about what size "a load" actually is, how tightly the logs are stacked, or, more crucially, what the wood's moisture content is. There is old wisdom in this simple measure.

What is a fair price?

As with any purchase, it is worth shopping around for reasonably priced firewood. You might think this would simply be a case of getting a few quotes, but with firewood you will rarely be comparing like with like – load size, species, and moisture content will vary. It is also worth remembering that when you buy firewood it is a package deal. Will the seller be careful to deliver the length, dryness, and species of log agreed? Will the logs be dumped at the end of your drive or stacked in your woodshed? Will delivery times be negotiable? Sometimes it is worth paying a little more to ensure the merchant is reliable and is sourcing his wood sustainably.

Heating is a major expense for most households and even though wood fires are a particularly beautiful fuel, you may be interested in going beyond just making sure that the price of your wood is fair and decide to make a cost comparison with the other heating fuels available. In recent years the price of heating fuels has varied significantly and it is good to know how wood compares. Helpfully the rising importance of wood fires has led to fuel comparison data being widely published (an example of which can be seen on page 217).

There are two stages to look at. Firstly, the cost of each unit of energy for the different fuels (wood, oil, coal, mains gas, LPG, and electricity) and then the efficiency of the burning system that delivers the heat. When burning wood the efficiency can vary enormously: an open fire is only likely to be around 15–20% efficient, while a wood-burning stove should be 60–80% efficient – some boilers and masonry stoves offer around 90% burning efficiency. Always remember low moisture content is extremely important in wood's burning efficiency. I have always found wood to be very competitive with other heating fuels.

Finding a supplier

The aim here, if possible, is to find a supplier that you trust and can build a long-term relationship with. For many people this security of supply is the key to enjoying a wood fires lifestyle. Once you have found a source of good-quality, sustainably produced firewood you can relax – the first and possibly the biggest hurdle has been cleared.

If you are new to wood burning I really would suggest that you devote reasonable time and effort to this issue – without a reliable supply of logs you have no fire. I do enjoy being alert to any opportunity to top up my woodpile, but would not enjoy the unsettling niggle of knowing that I don't quite have enough logs for the winter. The internet, local advertisements, and word of mouth should help you to find an initial supplier. Neighbours burning wood may be happy to help, or parents at the local school, pub, or village shop. Then try looking for the tree professionals in your area – obviously firewood merchants, but also timber merchants, sawmills, hauliers, tree surgeons, foresters, and chimney sweeps (you will need the services of a good chimney sweep soon anyway). The next route to finding a good log supply would be through trade associations, and this too could turn out to add a new and exciting dimension to your leisure time. Many woodland associations welcome new members, and attending open days and field visits may be a great way to improve your tree and log identification while making useful contacts.

Questions for your firewood supplier

Once you have found a good supplier, try and build trust without being naïve. Find an open, sympathetic, non-aggressive way of asking your questions – something like: "What is the mixture?", "When was this felled?", and "Where is it from?" It may be a rule of selling that customers tend to buy from people they like, but it works both ways. Those selling will always prefer to supply the customers they like too – a grumpy or confrontational customer will never get the best wood. During your conversation with the merchant you will learn what he has to offer and he will learn what you need, and can offer further suggestions or advice. Once the relationship is established, you could ask for a few of your favourite logs to scent your room, or if you require kindling you could enquire about this as firewood merchants often have it at a good price. It is important to maintain a level of quality control; after a delivery I usually select a few sample logs, split them, and check the moisture content. I may also take a few minutes to stack the load so that I can get a rough volume measurement and know what a "load" from this merchant really is.

The information you generally need to know from a wood seller would include:

— **Is he selling by weight or volume?**
— **What is the log size range and are they split?**
— **Are the logs hardwood, softwood, or a mixture?**
— **Are the logs green or seasoned?**
— **When was the timber felled?**
— **When were the logs split?**
— **What is the method of delivery?**
— **Does he deliver to your area?**
— **What is the price and does this include delivery?**
— **Is there a discount on bulk loads or spring/summer buying?**
— **Is there a discount on unseasoned, freshly felled wood?**
— **Where is the timber from and is this a sustainable source?**
— **Is the wood suitable for an open fire or is it prone to sparking?**

ALTERNATIVE FIREWOOD SOURCES & TYPES

Collecting your own wood

In this chapter I have focused primarily on buying firewood, but of course you may be able to gather some or all of the firewood you need yourself. Forgive me for stating the obvious, but I would like to open with the caveat that almost all the wood you see, anywhere, will belong to somebody. When I worked as a forest manager we would often find people loading the boots of their cars with logs. When challenged, their usual excuse was that they had "seen the logs lying around and assumed that nobody wanted them". Very often people will be only too pleased to let you tidy up some logs or branch-wood that they don't want, but to save any embarrassment, do ask first. A significant proportion of the wood that I burn each year comes from neighbours who are having tree work done, skips, beach driftwood, old fencing,

windblown twigs and branches, tree surgery off-cuts, and from contacts I make through belonging to a woodland charity and the Royal Forestry Society.

Slabwood and off-cuts

Sometimes you have to think laterally to get the firewood you need. Obviously the first lines of approach are to either buy your logs from a firewood merchant or to cut them yourself. Another line that may be worth following up, if there are any sawmills in your area, is to see if they sell slabwood and off-cuts for firewood. When large logs of timber are being sawn to make beams or planking the first stage is often to turn the round log into a square beam by cutting off four "slabs". If there is a lot of good timber still in the slabbing it may be reprocessed to make small pieces of sawn timber, such as battens. However, there will always be some residue of bark-covered slabwood. Some mills have developed a good market for the bark in garden products, but many, particularly the smaller mills, may be happy to sell it as firewood. Don't worry about burning the bark as it has very good heat values, although bark does have a much higher ash content than clean wood. Other places to try for off-cuts would be manufacturers of wooden furniture and fencing products. You will need to take some care if using off-cuts and slabbing as a major source of your firewood as it may be mainly coniferous softwood, which really does need long storage before it is good to burn.

Kindling

For no obvious reason I really love the word "kindling". These are the tiny sticks, the baby logs used to build the fire's nursery, the very beginnings of your fire. Without anthropomorphizing your kindling sticks too much, do give them special care. They should be well sized and absolutely dry as they are the key to lighting your fire quickly. Bought kindling can be quite expensive as there is a lot of work in chopping up the little pieces, so making your own may well be a good way to save a little money. It may be enough to simply gather fallen twigs during woodland walks, or perhaps driftwood from an old high-tide line. If not, put aside a few of your driest, knot-free logs for splitting. The most useful sized pieces are about 15–20cm (6–8in) long and 2.5–5cm (1–2in) in diameter. Dryness really is important and I keep a metal bin near my fire full of mixed kindling and pieces of bark that have fallen off my firewood logs. There is a temptation to make kindling from recovered wood, but this wood may contain preservatives, creosote, or old paint, and care should be taken not to burn this "dirty" wood. In fact, I use all sorts of things as kindling: orange peel dried on my hearth, the cardboard centre tubes of toilet and kitchen rolls, pine cones, and nutshells.

Engineered wood fuels

Of the three forms of engineered wood fuel, wood briquettes are the most like natural logs. They are made in much the same way as wood pellets, but are designed to be burned on either the open fire or wood stove. Briquettes offer the advantage of being a consistent, low moisture content (<10%), neatly packaged wood fuel, but they are generally a little more expensive than natural wood logs.

Wood pellets have certainly become more popular over the last few years and are now well established. This is probably because they are simple to use, offer a consistent, high-density, dry fuel which comes in clean and easy-to-use bags. However, much like briquettes, this is an expensive form of wood fuel – although this is partly balanced out by the high efficiency of most pellet-burning stoves. Some models of these pellet stoves are small, neat, and essentially designed as space heaters – something to consider if you have limited heating needs or wood storage space. The pellet stoves are also able to burn non-wood fuel pellets. Of course these stoves are still burning a wood fire, but somehow a very sanitized one. The fuel looks much like animal feed and is loaded into a storage hopper, which then feeds the flame – a steady little fireball. Burning pellets can be rather noisy.

The pellet system will certainly suit many people, but go into it with your eyes wide open. The fuel is a commodity you will always have to buy – no chance of picking pellets up during a woodland walk – and the stoves are fully automated, with microchip-controlled sensors designed to give you the perfect burn, so there is no room for you to develop your fire skills. The wood pellets are made from the sawdust and shavings from sawmills and other wood-using industries, such as furniture manufacturing. The raw woody material is hammered into a pulp and then forced at very high pressure through a die; the heated lignin within the wood plastifies and acts as the binding agent. As with many things in life, with pellets you will usually get what you pay for and the pellet quality may vary. This will depend upon the initial wood feedstock and manufacturing; watch out for pellets that crumble too easily and seem unusually dusty, or have bits of other material within the pellet.

Also, most pellet stoves need electricity to run as the air feed is fan-assisted and the fuel feed is often via a small auger. This will be significant if you are looking for a wood-fired heating system that will not be knocked out whenever there is a power cut.

Woodchip is generally associated with large-scale wood heating installations. If you have chosen this form of wood heating you will know that the fuel you need is an engineered woodchip – although it is not so much the actual woodchip that is the issue, but whether there is any unchipped debris mixed in with the chips. Splinters of wood, twigs, or stringy bark can form bridges in the fuel store where the auger is drawing the woodchips and cause the feed to fail. If your chips are an acceptable size and free of such debris they should be fine. This is definitely a time to liaise closely with your supplier until you have gained your own experience. The woodchip fuel is very bulky, needing much more storage area than logs or pellets. If you have bought a woodchip boiler, it is likely that at least initially you also have a woodchip supply contract. There are wood boiler systems that use round wood logs, but the sector of the market demanding the most heat, such as very large houses and schools, will usually have gone for a total woodchip heating system designed using professional architects and wood heating engineers.

CHAPTER FIVE

*

SEASONING LOGS

"By all means savour the delightful sound of sizzling bacon, but not the serpent's hiss of wet firewood."

—ROBERT RIPPENGAL

I sometimes feel that in the whole delightful art of making good wood fires, seasoning is the least understood aspect. I need to be coldly scientific for a few pages here as the subject is so important. Seasoning is drying, and drying is king in wood burning, but there is so much nonsense and bad information written and spoken about it.

Firewood is roughly 99% inflammable materials and water. To burn efficiently and cleanly, the fire needs to be hot and have oxygen freely available – there is no part of this unique exothermic chemical reaction that is helped by adding water. Adding water is in fact how we often put a fire out! It is not practical to remove all of the water from our firewood, but we should try to get out as much as possible. I don't want to be dull, but I do want to stamp on opinions like "firewood can be too dry", "it's fine to burn ash green", "adding a green log will keep your fire in overnight", and "logs should be seasoned for two years" – where did all this nonsense come from!? The general dryness benchmark for logs to be called "seasoned" is 25% moisture content, with the informed caveat that 20% is a better and easily achievable "ready-for-burning" target. The seasoning down to about 20% can be done at home in a decent woodshed, or by buying pre-seasoned or kiln-dried firewood.

In wood burning, the words "seasoning" and "drying" are often seen as synonymous and, fundamentally, they pretty much are; people refer to unseasoned wood as "wet" or "green", and ready-to-burn wood as "seasoned" or "dry" – but there is much more to it than this. It is very important for those who are routinely burning wood to truly enjoy and understand this subject. There are some very simple reasons why learning about effective seasoning is worthwhile – from global ethics to your bank balance. In a world trying to reduce the use of our declining fossil fuel reserves, wood is a readily available, renewable fuel resource that is almost carbon neutral; it is, therefore, precious and should be used wisely. Another ethical/legal point is that burning unseasoned wood and making unnecessary smoke is a cause of air pollution – something else the modern world is trying to reduce. The financial argument for seasoning is particularly persuasive, as burning your firewood badly wastes money. This is partly through fuel waste, as you will not fully benefit from the potential heat energy within your logs, and partly through avoidable repair bills, as burning unseasoned wood may cause material damage to your stove and chimney.

In this chapter we will look initially at how much time is really needed for acceptable seasoning, beyond simply adopting a broad one year/two year rule of thumb, and then how the time period is affected by the log size and tree species. After that, we will look in detail at what we are trying to achieve in terms of moisture content and, finally, we will consider the very essence of exactly why we are bothering to season the logs at all – that fabulously relevant aspect of thermodynamic physics, latent heat. Then, armed with all of this, you can approach the subject with confidence and discern good advice from the background noise of well-meaning waffle.

SEASONING TIME

For practical purposes, a rule of thumb would be that, for normal household split hardwood firewood, the logs should be seasoned through one full summer for use the following winter. Similar coniferous, softwood firewood logs are best given two summers. I would see these as sensible minimums. I have often read that firewood should be seasoned for "at least a year", but in many areas there is precious little drying through the winter months – it is the spring and summer weather that counts. Firewood felled in November will be little different in moisture content from wood felled the following March; both need the full summer season to be ready to burn the following winter.

Some people say that the time of year the trees are felled is not really significant, but I don't agree with this. It is traditional to try and fell trees in the winter when the sap is "down" (i.e. there is minimal flow); this is particularly important when coppicing. Other than it being better for the health of coppiced trees, there are two seasoning advantages to having winter-felled firewood. The first is simply that if you have less sap (which is primarily water) in the wood to start with, there is less to dry out during seasoning. The second advantage is more subtle and is to do with the nature of the sap itself, and the fact that tree sap is more than simply water.

Many years ago, while living in Scotland, I decided to see if I could make maple syrup from the huge Norway maple trees in a woodland behind my farmhouse. I was surprised to learn that my best sugary sap flows were in mid- to late February on cold nights, and most definitely still in winter. The spring sap flow of many trees begins in late winter and contains a high proportion of dissolved sugars and mineral salts. Firewood cut at this time contains a higher proportion of moisture and hygroscopic compounds that absorb moisture from the air; this to some degree adversely affects the ease with which logs can be dried and their subsequent capacity to re-absorb moisture. Another aesthetic point that I've noticed is that late winter- and spring-felled logs are more likely to develop unsightly fungal mould, which also inhibits a log's drying. There is a summer felling technique where the branches and leaves are left on for a few weeks, to suck out the sap. This has merit, but I would rather work in a cool and insect-free winter.

Shortening the seasoning process

The time needed to season a log so that it is ready-to-burn will vary with the size of the log. In practice, all this means is that if you need your wood to season more quickly, cut it into smaller pieces. Smaller, and particularly shorter, pieces of wood will have a greater surface area to volume, and when drying wood the larger the surface area to volume, the quicker moisture can escape. Also, as we are dealing with a material wrapped in a waterproof cover (the bark), breaking down the log size will increase the proportion of bark-free wood surface in contact with the drying air.

The old rule with oak in sawmilling is that beams and planks should be given one year of seasoning for every inch of thickness. I cut most of my oak into fairly small, wedge-shaped pieces with a maximum diameter of about 10cm (4in), which ensures that it is ready for me to burn the following winter. This addresses the oft-quoted rule that oak firewood in particular needs two years to season – large logs do need

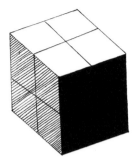

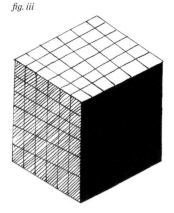

2.5cm (1in) cube
Surface area: 38.7cm² (6² in)
Volume: 16cm³ (1³ in)
Ratio: 6:1

5cm (2in) cube
Surface area: 155cm² (24² in)
Volume: 131cm³ (8³ in)
Ratio: 3:1

15cm (6in) cube
Surface area: 1,394cm² (216² in)
Volume: 3,540cm³ (216³ in)
Ratio: 1:1

Volume/surface area
Objects with a small volume have a proportionately large surface area. Fig. i. shows a small cube with a surface area proportionately six times greater than the large cube (fig. iii).

this length of time, but split them small and you shorten the seasoning process. The US Forest Products Laboratory and other research bodies have found that the need for "one- to two-year" seasoning times is a complete myth. Their experiments show that woods such as birch can be brought down to an acceptable moisture content in two to three months, if well split and stacked under cover with sufficient airflow. Similar claims have long been made for split firewood stacked in a Holz Hausen (see *Chapter 6*). Ben, a woodsman I cut firewood with, was the first person to show me how quickly wood can be air dried. He chuckles knowingly when discussing seasoning times. His logs are cut small and stacked roughly under the corrugated iron arch of an old Nissen hut – he swears they're ready to burn in only a few weeks.

So, the time your wood actually needs to season down to an acceptable moisture content will vary hugely depending on the size of the logs, the tree species they are from, and the quality of your woodshed. It's best to start out planning to give your wood plenty of time to season, but as you get to know the woods you are working with, you may find that you do not need to season it for as long as you expected.

Differing seasoning times

The species is important in determining an appropriate seasoning time. Trees with a dense wood structure, such as oak, elm, and hornbeam, will season much more slowly than trees in which the wood is not dense, such as alder, birch, and willow. The cell structure of the different tree species will also have an effect on drying time.

A broad guide to the time needed to season the different hardwood tree species is:
- **Very slow-drying:** English and holm oak, English elm, hornbeam, yew, and box
- **Fairly slow-drying:** beech, robinia, fruitwoods (cherry, apple, pear, etc), hawthorn, blackthorn, holly, horse and sweet chestnut, red oak, laburnum, rowan
- **Fairly quick-drying:** ash, birch, lime, wych elm, sycamore, maples, planes, hickory, hazel, walnut
- **Very quick-drying:** willow, poplar, alder

Most trees grow rapidly in springtime and growth then slows through the summer months. In some tree species this annual cycle of spring/summer growth forms distinctive bands radiating from the centre of the tree; these annual rings can be used to age a tree. "Ring porous" species such as oak, sweet chestnut, ash, elm, robinia, hickory, and mulberry have less ability to dry laterally than "diffuse porous" species such as beech, sycamore, maples, plane, alder, willow, birch, holly, and poplar. However, despite these general rules nature rarely sits neatly within man-made definitions and many tree species are in fact somewhere between being truly ring or diffuse porous – for instance, cherry and walnut are often considered to be in this intermediate class. Conifers have an entirely different cell structure and in general take longer to dry than most hardwoods.

Using wood from conifers

The conifers are in general seen as the second-class citizens of the firewood world. There are many good reasons for this prejudice as they are slow-drying, full of resins, apt to pop and throw sparks, and, not being very dense, take up a disproportionate amount of room in the wood store. However, they are very widely planted and should not be rejected out of hand. For the open fire I would say that if you must burn them, you'll need a good fireguard. However, the increasingly widespread use of wood-burning stoves with properly lined

chimneys should result in a change in our attitude towards conifers. I recently looked at a pile of huge old Scots pine trunks stacked on a forest road with a view to buying them. The logs had been piled up as part of a large felling operation and then never sold – they had lain untouched for at least seven years. Most of the bark had fallen off and the wood was bleached white in the sun. The logs in contact with the ground were partly rotten, but the upper logs were beautifully dried and would have made very acceptable firewood. For my wood burner I would be happy to take old cedar, pine, larch, and cypress; I would be more reluctant to take spruce, hemlock, redwood, and firs – see *Chapter 2* for more information on burning conifer wood.

I visited a tiny village high in the mountains of western Crete one autumn and, walking between the houses, was delighted to find the air slightly scented with the aromatic, resiny smell of cypress wood. In the shelter of this high valley there were plenty of hardwood trees, mostly olives, planes, and mulberry, but I was surprised to find six woodpiles made up purely of cypress logs. The villagers made the best of what they had, blessed with the searing heat of a full Aegean summer to dry their wood. As I seem unable to separate wood fires and food – both soothing pleasures to the senses – I'd like to mention the snack I had there: smoked pork, honey pastries, and a curious tea, woodsy with the taste of mountain herbs. The pork was superb – tough and bursting with smoky flavours. The honey was sublime, best-ever-tasted, out-of-this-world good – clear, thick, amber-coloured, and infused with the flowers of every tough little plant the bees could find on the parched limestone cliffs.

Before leaving tree species I would offer one final, overarching thought. Learning all about the trees and your favourite types of wood is thoroughly enjoyable. But do keep in mind the simple, boring fact that, weight for weight, there is very little difference in the heat energy available from the different tree types. It's not the type of tree but the moisture content that will make all the difference in the world to your fires. This point was brilliantly captured in a conversation with German friends of mine who live in a huge wood-heated house in the Black Forest. I asked them what was the one thing that they wanted from their firewood. Without hesitation the wife replied, 'For him to dry it properly!'

A good-natured argument ensued as the husband defended the quality and dryness of the logs he brought her. They understood that proper seasoning is one of the great secrets of good fires, and you should be very wary of any advice to the contrary.

MOISTURE CONTENT

....

This really is what seasoning is all about. Reducing the moisture content of your firewood down to a practical point at which the wood will readily burn. I say a practical point as obviously the drier the wood is, the better it will burn, but with natural seasoning you can only bring the wood logs down to the ambient moisture content of their surroundings.

It is worth taking a moment to look at how the moisture is held within wood, as this helps to understand how the logs dry out. To a very large extent, wood is just a mass of tiny long tubes running along the length of the tree, and the moisture within a log exists in two distinct forms: as liquid water within the cell cavities and as vapour/molecular water within the actual cell walls. The water within the cell cavities is often called "free moisture" and is the

Wood vessel moisture
Wet wood fibres in green wood (fig. iv); fibres at saturation point (fig. v), and (fig. vi) much drier wood fibres in seasoned wood.

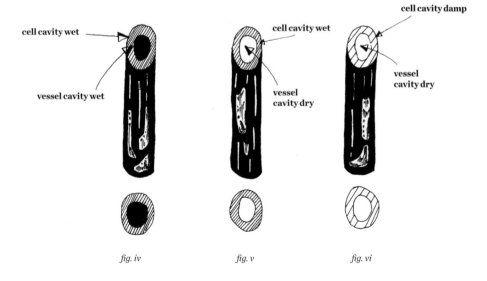

cell cavity wet

vessel cavity wet

cell cavity wet

vessel cavity dry

cell cavity damp

vessel cavity dry

fig. iv   *fig. v*   *fig. vi*

first and easiest to get rid of during seasoning. When the average hardwood has been dried to a moisture content of around 30% most of this free moisture has gone, but the moisture within the cell walls still remains – this is called the "fibre saturation point". It takes longer for the moisture that is actually trapped within the cell walls to leave the log. As the log dries below the fibre saturation point, the wood will begin to shrink and crack. Confusingly, in the world of wood fuel there are three different ways in which the moisture content of wood is determined. The most common, and to my mind most sensible, is to take the gross weight of the log and determine what proportion is wood and what proportion is moisture. With this method, if a log weighing 1kg (2.2lb) has 600g (1.3lb) of wood fibre and 400g (0.9lb) of water, it would be said to have a 40% moisture content.

The moisture content of freshly felled trees in mid-winter varies from around 35% in ash up to 50–60% in trees like alder, willow, and elm. In general it is enough to assume that freshly felled trees will have a moisture content in the region of 50%. Most experienced people who burn firewood try to dry their logs to something like 20% moisture content. In late February I thoroughly tested the moisture content of logs in my woodshed. This seemed a sensible time to do it as the wood had had about four months of winter to lose their peak end-of-summer dryness, and this was the time I most needed my fires. To my surprise, there was some variation between the species. (For accuracy, I did the test with two moisture meters, both with new batteries.) My average results were: pine 13%, alder 14%, birch 15%, oak 16%, willow 17%, and beech 19%. I found no difference between my one- and two-year-old logs. To help the drying, I do split even small logs to break the bark and my logs are never long, averaging about 25cm (10in), as logs dry mostly through the ends. Long logs, not yet cross-cut and still "in-the-round", dry very slowly, if at all. I am in the habit of stacking a few logs (safely) close to my wood burner for a final week of drying beside the stove, so I tested these too. They were beech and now showed a huge dryness variation within each log. The log end facing the stove had dried to around 5%, but this was only for the first 5cm (2in) of the log. The log end facing away from the stove and most of the rest of the log seemed little changed, at around 19%.

Using a moisture meter

The moisture variation between types of wood and indeed within individual logs means that some diligence is needed when using a moisture meter if you are to make a meaningful assessment of your firewood. You obviously can't sample the moisture content of all of your logs and so it is necessary to take an appropriate sample. When considering an assessment of the moisture content of a heap of firewood I would suggest sampling something like 12–20 of the logs. If the results vary hugely then you may need to take a larger sample. In practice, if you are finding a range of readings from around 20% to, say, greater than 35% then this probably means somebody is trying to palm you off with some unseasoned wood hidden among the seasoned logs.

How moisture meters work
Electricity flows readily through a damp environment and poorly through a dry one; moisture meters use this principle to test material for electrical conductivity and, thus, for moisture content. They work by having two sharp prongs that are pressed into the wood sample and then pass a small current through it. The resistance (dryness) in the circuit is measured and converted by the device into a moisture content reading. Different readings from different areas of the same log are common.

The second point to note in your diligent use of a moisture meter is that it is often misleading to take your readings from the surface of the log. I had some oak that had been split and stacked in my woodshed for roughly three summer months. The ends and outer faces of these logs gave me a moisture content reading of 18–20%, but when split the freshly cut inner surfaces indicated that the majority of the wood had a moisture content of around 35%. These logs were far from ready to burn and the half-hearted use of a moisture meter on these logs would have been very misleading indeed. The opposite may also be true – perfectly good logs that have been out in rain or drizzle recently would give a wet reading on the log surface, when in fact the logs are fine and would be dry again in a very short time. Also note that the heartwood and sapwood within a log will probably have different moisture contents. So, split the test logs and take your readings from across the freshly cut surface.

But do keep your use of moisture meters in perspective, wood fires are as much an art as a science and all you need is an approximate moisture content figure to guide you. Also, I have two meters and they give slightly different readings!

LATENT HEAT

This is something that I struggled to understand at school, but with wood burning the latent heat phenomenon is very important. It is generally understood that dry wood burns better than wet wood and, therefore, seasoning must be a good thing. But I think that to really grasp the hopeless waste of heat energy when trying to burn damp wood, it is worth having a quick look at latent heat.

Matter usually exists in three basic states: solid, liquid, and gas (we won't bother with the fourth state: plasma). Let's say first that we want to raise the temperature of a body of water by 1°C (1.8°F) and to do this we use one unit of heat energy – delightfully, this is called "sensible" heat. Now let's say we want to melt enough ice to form the same body of water – to change the state from solid into liquid. This would take around 80 of the same units of heat energy and the temperature would still be 0°C (32°F) – this is "latent" heat – the heat energy needed to change the state of matter, but not its temperature. The heat energy absorbed in this endothermic reaction goes into weakening the intermolecular forces and bonds to the point where the matter changes state. Water has particularly strong bonds and therefore requires even greater heat energy to change into its gaseous state – steam.

It takes a staggering 540 units of (latent) heat energy to turn the watery sap into steam – a truly massive waste of the heat energy available in a log. It boils down to seasoned firewood having more or less double the potential heat energy of green/unseasoned firewood. So anyone burning their wood green is wasting roughly half of their logs. It's not just that damp wood burns badly, but using the heat energy within a log to unnecessarily drive out the moisture is an appalling waste. Trying to burn unseasoned wood is like trying to get warm in wet clothes.

CHAPTER SIX

*

THE
WOOD STORE

"Every man looks at his wood-pile
with a kind of affection."
—HENRY DAVID THOREAU

I have seen nasty things in woodsheds and this is a great pity as it is very easy to store your precious firewood properly. I think it helps greatly to develop a good attitude towards your woodshed; it is not some sort of slightly damp attic, into which you can throw things you feel you may need at some indeterminate point in the future.

Neither should it be seen as a necessary inconvenience, an annoying waste of garden space, or an eyesore to be hidden somewhere distant among compost heaps and old sheds. The woodshed is a larder of joy, not to be hidden away or given scant regard. I say "larder" and I mean it. Just think what this store is actually holding – not the rough, raw lumps of beech, birch, oak, apple, and ash, but the potential to provide you with radiant, sun-like warmth, ever-changing, flickering light, beauty, aromatic, smoke-scented rooms, and toasted food! We all store our food to the best of our ability – nobody accepts mice, mites, mould, or damp ruining stored food, and so it should be with your firewood. Aim not for acceptable wood storage, but for the very best you can provide – as you do with your food.

In many places there is a long tradition of burning coal, and storing coal is easy and quite different – it is fine to tip coal into a damp bunker and forget about it. But wood is elite, the thoroughbred, and must be well housed. I rather like one of Oscar Wilde's observations in *The Picture of Dorian Gray*: "The one advantage of having coal [mines] was that it enabled a Gentleman to afford the decency of burning wood on his own hearth." So here we are, affording the decency of wood fires on our own hearths, alongside Wilde's Gentlemen.

This chapter firstly works through the technical points to be considered in creating the best storage for your wood and then looks at some of the wider, more subtle ways in which you can enhance the woodshed. It was Ada Doom, the matriarch in Stella Gibbons' book *Cold Comfort Farm*, who was constantly haunted by having once seen "something nasty in the woodshed"; let's make sure there's nothing nasty in yours.

Finally I look at some of the alternative ways of storing and drying your wood. I remain unconvinced whether a solar kiln is in fact of any real practical benefit to the vast majority of wood-burning homes and I discuss the pros and cons of these.

WIND, RAIN, SUN & SHADE

The wind

Before we look at the different types of shelter and coverings that are acceptable as "woodsheds", let us first look at what exactly we are trying to achieve in order to store our firewood well. Most people immediately assume that we simply want to keep the rain away from the logs and, while this is obviously a major factor, it is not actually the most important thing. Perhaps think for a moment about the dynamics of drying a line of washing. Only in the very wettest climates do people actually hang their washing under some form of roof. I know the washing is only expected to be out for a few hours, but I think it illustrates the point well – even if there are occasional rain showers it is the airflow, the wind, that dries the clothes the most quickly. So our first consideration is wind: the flow of air through the logs and above and below them. Even in the driest, most waterproof shed, a load of damp logs will be very slow to dry if the door is closed and there is no airflow. So use the wind as much as possible. When creating the wood store try to use any naturally windy spots, perhaps where the breeze is funnelled to produce a windy gap or corner, and consider this also when positioning heaps of pre-drying logs. In areas with a dry climate it is usually enough to let the wind alone dry your firewood and not to rely on the sun.

The rain

The second climatic consideration is rainfall, or perhaps more accurately general precipitation, and the wetter the area you live in the more important this is. Although that is not quite true, as what really matters is the way in which the rain falls, not how much. Your quota of annual precipitation falling as infrequent heavy showers is much easier to deal with than if you commonly have days or weeks of continuous light drizzle. Rather like the washing described earlier, a pile of dry logs can take a heavy rain shower and soon be dry again if there is a good airflow around them, however this is not the case in regions with prolonged periods of light rain and drizzle.

In a climate like this, clearly your firewood will need very good protection from the rain and drizzle, but while the wood must obviously be under cover, I would still argue that good airflow is the all-important factor. Also, in most areas the rain tends to fall disproportionately from one direction: the direction of the prevailing wind. If you do not know what this is, you can quickly find out by asking local people. This local knowledge is important as the direction of these moisture-laden winds will vary hugely in different regions and countries. In Britain and Ireland it tends to be winds from the south-west to north-west (through west) that carry the most rain. Knowing this, you can then set about protecting your wood from these rain-laden winds, while allowing abundant airflow from the opposite direction. So, it is important to consider how best to protect your firewood from the rain, but you should be clever about it and not build more, or enclose the logs more than you need to; as I keep emphasizing, good airflow is vital.

The sun
It is a good thing to have strong summer sunshine beating down upon your firewood, but in practice this will always be difficult. Also, of course, the sunlight can only ever reach the outer wall and top layer of the logs. Nevertheless, if it is convenient to you and you are planning that the primary rain protection should be from the south, west, and north, then it may be reasonable for your woodshed to be open to the east and therefore bathed in summer's morning sunlight. I find that direct sunlight falling on the logs helps the bark to develop fractures and break up, which then allows moisture to escape radially from the log – most intact bark is more or less waterproof. The sunlight will also create tiny fractures in the cut end of the log and this may also speed up drying although, in my experience, this end-drying can sometimes shrink the tiny vessels that run lineally along the log and limit drying. On balance, sun cracking of the bark is a good thing, although overall direct sunlight only offers a limited advantage and is generally a fairly minor consideration with woodsheds.

Mature trees and their shade

Trees can be either a help or a hindrance in the siting of a woodshed and the storage of firewood generally. As anybody with experience of camping will tell you, good practice is never to pitch a tent beneath trees, primarily because the trees will continue to drip water long after a rain shower has finished. This constant dripping, together with the shade, creates a microclimate beneath a tree, woodland, or forest canopy that is generally damper than the microclimate in open country.

Having said that, those few tree species with an extremely dense canopy above and no restriction to the airflow below can be a great help with wood storage. I'm sure that it would be possible to create a 100% natural, living woodshed through the judicious use of trees. In the dead of winter it may be worth checking the ground around any suitable conifers that you have; if the ground beneath them is dry, you may have discovered a small, free wood store.

In my twenties I lived in a cottage in the heart of Rendlesham Forest in Suffolk, where I burned wood, kept chickens, and often went beachcombing. It was watching my chickens that first gave me the idea of using some of my standing trees as wood stores. The hens roamed freely around my entire garden and enjoyed the occasional dust bath. I was surprised to note that their best four-season dust bath was under three Lawson cypress trees. These are conifers that form tall, spire-like trees with dense, scaly foliage. When open grown, the western red cedar and Leyland cypress form a very similar-shaped crown to the Lawson cypress. Once I realized that the foliage of these trees was intercepting virtually all rainfall and the ground immediately around the base of the trunk was dust dry, the chickens had to make way for the log store. From then on I stacked split logs in a circular column around each of the three Lawson cypress trunks, where they seasoned beautifully.

SITING & SETTING UP
YOUR WOODSHED

There is a golden rule in woodland management not to move any wood more times than you have to; unnecessary moving is called "double handling". This is a key consideration in choosing where to site your woodshed. A perfect site will allow the wood logs to be delivered easily, have room for you to do any further splitting, and be near enough to the house to make bringing firewood into the home easy and convenient – especially in winter. In short, the perfect site means little or no double handling. In addition to siting the woodshed to make delivery and handling easy, there is a range of other issues that may need to be considered.

Pest avoidance

A minor consideration in the British Isles, but a major consideration in other parts of the world, is pests. Thankfully I have never had to consider the potential problems caused by carpenter ants or termites; where these pests are endemic it is always wise to have your woodshed sited well away from your home. But it is worth just thinking through the possibility that you may get rats nesting under the wood pile, and that wood-boring insects will occasionally breed in your logs and later emerge in their thousands. The woodshed is also a very good place for wasps nests, perhaps even hornets. My last woodshed had the hugely more agreeable pest of rabbits nesting under it. As the woodshed was adjacent to the orchard I really didn't mind, but my neighbour was rightly proud of her immaculate vegetable garden and had a much stronger view on the matter. For more information on potential residents in your log store, see Attracting Desirable Wildlife on page 124.

In this chapter we look carefully at the wider and more general considerations to be thought through when first planning how you intend to store an adequate supply of wood fuel. A rough summary of this would be:

— **Your wood store must be big enough, well ventilated, and have a sound roof and good protection from the prevailing rain-laden winds.**
— **The flooring must be free-draining and if possible also allow some airflow under the logs.**
— **The building should be robust as it will take some rough treatment.**
— **The location should be suitable for the easy delivery and collection of logs.**
— **The store's design should blend in and "belong" in its surroundings.**
— **Finally, your shed should be something that you are proud of and enjoy spending time with. This is a special building – a gorgeous, rustic larder of winter warmth and joy.**

Of course there are almost as many types of woodshed as there are people who burn wood, each store being adapted to suit the owner's individual property and firewood needs. But I think the woodshed illustrated opposite comes close to deserving the moniker of an "ideal" woodshed. It is built of stout timber, with a watertight roof and one solid wall facing the worst of the weather. The two side walls are slatted with good planks – not, as I have seen, flimsy feather-edge fencing boards. In the fashion of old cart sheds it has one aspect completely open, allowing total freedom of working when handling the wood logs. Finally, it has 15cm (6in) of 2.5cm (1in) loose stone over a permeable woven barrier, which extends out in front of the open side to make a dry and mud-free chopping area. Access is good; a trailer of logs can easily be parked alongside the shed and the wood thrown directly into it.

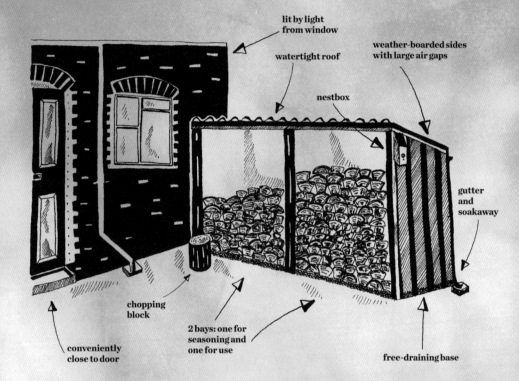

lit by light
from window

watertight roof

weather-boarded sides
with large air gaps

nestbox

gutter
and
soakaway

chopping
block

conveniently
close to door

2 bays: one for
seasoning and
one for use

free-draining base

The ideal woodshed as shown here is perhaps too open for those in a relatively wet climate. However, the simple addition of a tough tarpaulin panel for each of the two open bays would remedy this and adapt the woodshed for prolonged rainfall. If absolutely necessary it is also possible to put outer cover planking to prevent driving rain getting into the shed via the air gaps in the slatted end walls. The design has two bays: one for seasoning and wood capital storage, and the other for the split, dry logs that are ready to use – the current account. Of course in all areas there will be some rainfall and careful thought needs to be given to what happens to the water that drains off the roof. In an area of relatively low rainfall with a freely draining soil, a roof that slopes away from the open side and drains without the aid of guttering is more than adequate.

Using other buildings and walls

Another major consideration in the siting and storage of firewood is the aesthetic and practical opportunities offered by walls, other buildings, and the general style of your garden. Whatever you are planning to use as the store for your wood it must fit in with the look and feel of the rest of the garden. It is wise to utilize anything that will save you time and money and help the wood store to look like it belongs or has always been there. In general I would be reluctant to adopt an existing building as a wood store, however convenient that may be, unless I was absolutely confident that the building was dry and would allow an adequate through flow of air – a cart shed perhaps, or by having one door open on either side of the building for large parts of the year. Remember, you are creating a perfect "larder" for your firewood; a makeshift, almost-good-enough version won't do.

Multiple woodsheds

I keep using the singular term "woodshed", but in fact, although this rarely happens, it is completely reasonable to have more than one woodshed. I have seen people using two or three wood stores for three primary reasons: to work a deposit and current account system of firewood storage, to have one store for dry logs and one for seasoning, and lastly as they simply do not want one big store and are utilizing several nooks and crannies around the home and garden to season and store their wood fuel. A further advantage of any of this is that it gets you away from the "dry logs under the wet logs" syndrome that often plagues people working with one wood store. Let's look quickly at each of these multiple storage systems.

The deposit and current account system seems to work really well. The deposit account woodshed is a store that may hold two or even three years' supply of wood, all split and seasoning and not necessarily close to the house. This is your store of wood capital, topped up whenever the opportunity arises and preventing the need to buy when seasoned firewood is expensive. With this capital reserve you can then relax, knowing that you can see out the next winter or two regardless of what firewood comes your way, and can ignore the local seasonal price variations. The current account woodshed is close and convenient to the home and may only hold enough wood for a month or two. It is small, neat, and stacked with fully seasoned, split, and ready-to-burn logs. The most obvious example of this that one sees in winter, particularly in Europe and in North America, is where logs are stored ready for use under the eaves of a house.

The system of having one store for dry logs and one for seasoning also serves to build up a decent store of wood capital, but this time the two sheds are roughly the same size, each containing the wood logs needed for one year or more, and you use them alternately. I like this system too, as apart from affording you a good reserve of capital there are two other advantages. The first is that you do not need to double handle any wood – the split, damp logs are thrown into the seasoning shed and not touched again until a year later, when you pick up the dry, fully seasoned logs to take them to the fireside wood basket. The second advantage is that it allows you to easily accumulate green firewood at any point during the year, as this all goes into the seasoning, not-ready-for-use woodshed. With only one wood store, adding freshly cut wood to your seasoned stack involves either pulling out an area of dry logs to put the fresh logs underneath and then re-stacking, or creating a temporary store. It is a good idea to plan your wood storage system so that you have the capacity to easily take in free or cheap wood, whenever the opportunity arises.

I confess to not being very keen on "nooks and crannies" firewood storage. I have seen homes in which the firewood is kept in a range of different places: stacked against rainwater down-pipes, in the narrow, damp gap between a shed and a fence, in a recess beside a bay window, even in damp cellars. On the face of it this system utilizes otherwise unused areas and should give you the advantages of the two systems above – building a store of wood capital and keeping seasoned and unseasoned wood separate – however what actually seems to happen is that most of the nooks and crannies used are convenient, but are not really very good places to store wood. They can be damp, catch and hold leaves, have weeds such as stinging nettles growing in them, and are seldom adequately roofed or ventilated – in short, they are often makeshift and clumsy. Also, the person who is squirrelling away the logs may have a tight grip on what is ready to use and what is not, but if there are little stacks all over the place it makes it very hard for anyone else to know which logs are ready for burning. Your firewood is really important, valuable, and hard-won; you wouldn't store your heating oil in makeshift containers all over the place, so it is wise to show your firewood the same respect. Plan your wood storage purposefully and work to a plan.

Lean-to woodsheds

Having carefully looked at a robust, purpose-built, stand-alone woodshed, let's now look at other perfectly reasonable buildings and systems for storing your firewood. The first would be where you have the opportunity to utilize an existing wall. This building would be fundamentally the same as the "ideal" woodshed, but obviously with one free wall it is cheaper to construct and, with care, more readily may appear timeless, embedded, and belonging to the location. Fairly small ready-made woodsheds of this type are now widely available at larger garden centres and do-it-yourself stores. There are two interesting points to note with the good lean-to woodshed illustrated on page 117. The first is the aesthetic aspect of design in that, as this building sits beside an 18th-century cottage, the frontal upright supports are of weathered, reclaimed English oak posts, allowing the new building to sit comfortably beside the older one.

The second point is the more prosaic one of drainage. In this example the roof is sloping backwards towards the wall, but as the wall is a legal boundary, the water is draining onto the woodshed owner's property and down inside the back of the woodshed. In this case guttering is essential or there is no hope of properly seasoning the wood at the back of this lean-to. A low lean-to woodshed should have a roof sloping towards the inner wall, while with a higher building the way the roof slopes is largely a matter of personal choice. However, if the roof is able to slope outwards this will provide the logs with a little more protection from the rain.

Stacking logs for seasoning

I love the fact that the whole subject of wood fires is so accommodating. You can do just about everything fairly badly and get by, but how much more rewarding it is to learn to do things well. Like Thoreau, let's make sure we always look at our woodpile with a kind of affection. In the stacking of firewood be mindful of the logs' requirements and your own. The logs' needs are the now familiar mantra of "rain off and air through". You need your wood stack to also be stable and make the very best use of the space available. Don't assume stability – build your stack carefully, as wood logs move as they dry out and shrink.

If you are blessed with much more wood storage room than you actually need, then you can just tip your logs into a large, low heap. If, however, you have a limited storage area, the need to be tidy and make the best use of your available space is paramount. The "space available: need to stack neatly" ratio gives us three broad options: 1) no stacking at all (the logs just rest in a heap as they fell); 2) the loose wood is retained by a perimeter log wall; and 3) all logs in the woodpile are stacked neatly. There is not really a best practice to recommend here – you must simply stack your wood as the available space and your inclination dictates, but unless you have a cart shed or barn, you probably won't have the room to just throw your wood into a rough heap. Rough heaps have a greater air space, will tend to form a low pyramid, and may well fall on you as you collect logs.

Rough heap

Logs placed roughly in a heap upon accepting delivery.

fig. i

fig. ii

Stacking behind wall
Logs piled up roughly behind a log retaining wall (fig. i). All logs are stacked neatly behind the log retaining wall, to use the least air space (fig. ii).

Logs are generally more or less round and this is unhelpful when trying to build a stable log wall, especially one with corners. In *Chapter 7* we look at why it is best to chop up most of your wood, and one of the reasons is for stacking. Once you have some skill with an axe you can cut a proportion of your logs into square or rectangular blocks, and this is hugely helpful in the subsequent stacking. A log wall built to retain the rest of the unstacked firewood should be designed with the following points in mind. The half-round logs should be placed bark upwards to shed rain more easily and prevent any moisture build-up in the lowest part of the gutter-shaped and waterproof bark. The front log wall must be tied into the logs behind it in some way – to prevent the wall from collapsing outwards. I do this by using an occasional long log, which protrudes back from the front wall into the wood stack behind, or, if the wood pile is to be high and I want to be assured of stability, I will use short planks of wood to tie the stack together. My corners are carefully constructed, log-cabin style, using the rectangular chunks I split and, these corners are tied into the adjacent walls.

Another tiny subtlety I find helpful when making a large wood stack is to try and build in some further stability by making rough internal walls within the heap itself – particularly if it is going to be higher than 1.5m (5ft). Just occasionally I have to go out on a dark winter's evening and grope about for the logs within my unlit woodshed, and I don't want a bruising mini-avalanche tumbling at me out of the gloom as the stack suddenly gives way.

Lighting in the wood store

Throughout winter I try and bring in all the firewood I need for the week during the hours of daylight – during the shortest days, this only gives me the weekends. It is obviously high time I adopted some form of lighting in my woodshed. If you decide to install an electric light, take care that everything used is approved for outdoor use and that the switch and wires are waterproof. I would also suggest some guarding for the light bulb itself. My daughter Daisy has fitted a solar-powered LED garden light, with a movement sensor, to her woodshed, which works really well. When collecting logs in the dark you only really need just enough light to see by, and a dim light will help you to keep some night vision for the walk back to the house.

MAKING THE WOOD STORE BEAUTIFUL

The woodshed is by necessity a practical building, serving the twin masters of wood seasoning and storage. Nevertheless, there is absolutely no reason why you should not strive to also make it something lovely. This has to be a good thing, as long as your ideas for improving the aesthetics do not compromise the basic functions of the building.

Flowers and trees

The hard exterior of the building can be softened by growing climbing plants such as honeysuckle, jasmine, or roses, up the sides. I would avoid any plant with dense foliage that may in time restrict airflow, such as ivy. I would also avoid those plants with legendary vigour, such as Virginia creeper and Russian vine. They may soften the building very quickly, but you will pay for it later through the frequent pruning needed to keep the rampant growth in check. If you do like the idea of growing something pretty up the sides of your woodshed, bear in mind that the entire plant may have to be cut back from time to time in order to carry out maintenance on the building. An easier alternative may be to plant a large bush or small tree close to the building – close enough to give the required screening, but still far enough away to allow for routine maintenance.

Creative wood stack design

If you have the time and inclination, you can make the wood stack itself beautiful. This could be in the manner of a well-made dry-stone wall – displaying clinical neatness, all logs stacked with orderly precision.

Creating a recessed feature is another option and something I've seen in Switzerland, the land of window boxes and hanging baskets, where small frames or arches are built into the front of the wood stack among the logs. Pots of trailing flowering plants are then placed in these little recesses and left to cascade down across the logs below them.

Not long ago I bought in a couple of tonnes of firewood that contained a high proportion of bog oak. Now bog oak is dark, clay blue/black, and strikingly different from the much lighter ash and beech logs I was intending to stack with it. Once all the logs were split it took very little effort and was a lot of fun, to build a striking pattern into the face of the woodpile. As this was the seasoning shed, it made me smile every time I passed it.

Attracting desirable wildlife

Earlier in this chapter I talked very briefly about pest avoidance, watching out for undesirable wildlife in the woodshed – rats, termites, etc – but what about desirable wildlife? Wild birds probably give more joy to people than most other forms of wildlife and, while spiders, insects, snails, toads, and the occasional mouse may use your woodshed, very little is needed to encourage them. Many woodsheds have open sides and are infrequently disturbed during the main nesting season, therefore they are ideal as a nest location for many bird species. A nesting box for blue tits can easily be mounted on an east-facing end wall, or a box designed for robins or spotted flycatchers could be discreetly embedded in the building. In time, swallows may explore the rafters and find a snug shelf on which to build their nest. If not, you can always build them one. Of course with encouraging these birds comes a responsibility not to stack the wood in such a way as would give a cat easy access to the fledglings. If you enjoy the idea of making your woodshed wildlife-friendly, you could also look into providing boxes for bats, or providing bumblebee roosts. Anybody can create a simple, dry wood store, but with imagination you can make it so much more than that.

ALTERNATIVE WAYS TO STORE WOOD

Simple air-drying stacks

This chapter has so far focused on storing and seasoning firewood within some form of a building, or, at the very least, under a roof. However, it is a common practice in many parts of the world to achieve storage and a degree of initial drying by leaving the logs in a simple open stack. Here I am not talking about large timber stacks beside forest roads, where the wood is simply awaiting sale and removal. Of course, the drier the climate the more effective the open stack technique will be. I first saw this in a woodland in southern France one spring where, after a thinning, neat piles of cleft firewood logs had been leaned against the remaining standing trees to dry. More surprisingly, I have also seen drying stacks created in the west of Scotland – here the logs were in open country, criss-crossed and built into roughly 2m (6.5ft) columns, with the airflow maximized. There is obviously the important question of stability to consider when building something like this; it is not somewhere I would want to see children playing. Although here we are specifically recording that storage and drying can be achieved in an open stack, why not add a roof?

Beehive stacks

Many years ago, on a family holiday to the west coast of Ireland, I was intrigued to learn about the immense skill used by ancient peoples in constructing "beehive" houses. These dwellings, made without wood or mortar, use only the local, randomly shaped rocks such as you would see in any upland dry-stone wall. The beehive's walls arch gently upward to form an igloo-like roof. Now this alone demonstrates an absolute mastery of building without mortar, but I noticed another thing too. The individual rocks that made up the walls had been carefully placed so as to have a slight outwards slope – these walls were designed to naturally shed the rain. I was hugely impressed, and from then on determined that I would strive to construct the outer walls of my log piles in a similar, rain-shedding manner.

It is fairly cheap and easy to provide some protection for the wood stack from the snow and rain, using boards or sheeting, and the drying result should be greatly improved.

A favourite form of open stack is the Holz Hausen. If we look at the shape of traditional round haystacks, or the ancient beehive stone buildings of Ireland, on reflection it seems pretty obvious that this must be an efficient shape to stack wood. Making a Holz Hausen open woodpile is an efficient use of space and allows sun and air onto your wood. Most are built with the clever little trick of having a marked pole down the centre to let you know when the wood is seasoned – the mark shows when the stack has shrunk by up to 20% and so should be dry enough to burn. I have no doubt that these beehive stacks are a good idea, but I do have reservations about the astonishing claims sometimes made for them. If you fancy making one and look into it, you may read that they dry wood very much quicker than any normal wood stack through a central chimney effect; I'm yet to be convinced of this. But I do believe that this stacking method is a very worthwhile technique that should have a place in our wood storage.

Long log storage

Occasionally you may be in the position of needing to store your firewood in long logs, perhaps even several metres in length. Clearly the opportunities for drying to occur linearly through a long log are very limited, so you must maximize the radial drying. Tree bark is almost waterproof and must not be left intact or the wood simply cannot dry. An easy way of creating an initial fracture in the bark is to run a chainsaw lightly along the whole length of the log; the bark will then hopefully crack in the sun and soon begin to fall off. Where I have seen this done the larger logs had at least three of these chainsaw cuts along them. If a chainsaw is not available, a garden spade could be used as a peeling iron and a strip of bark removed along the length of each log. Where timber has been mechanically harvested, the bark will usually be punctured and torn by the feed rollers; in this case there is no need to further damage the bark to facilitate some drying.If these long logs are to be stored horizontally, it is important that they are rested on bearers to allow a free flow of air beneath them and are not

touching the ground. If the logs are to be stored vertically, for instance leaning against a building, then a way must be found to prevent the log end from absorbing water from the ground. In this case bearers are not really an option and pallets, steel weld mesh, or rough stone are usually used as a base.

Tarpaulins

Tarpaulins are marvellous things – simple, cheap, and waterproof – yet all too often a disaster in terms of wood storage. I think it is the very convenience of a tarpaulin cover that leads people to be casual with them and often use one as a quick fix: the dreaded temporary shelter. Perhaps it is the time pressure of modern life; when a load of wood comes in, and there really is no time to deal with it properly, the easy answer is to leave it where it is and lay a tarpaulin over the pile. The tarpaulin is not a roof, it's a cloak which gently moulds itself to the shape of the wood pile. The weighted edges seal the base, the sun beats down and you have a fabulous, damp hothouse, ideal for fungal propagation. In the rush to keep off rain, the crucial element of airflow was forgotten.

The above is all very negative and of course tarpaulins can be used well, but I have seen more good firewood ruined under tarpaulin than anywhere else. So how should we use them? The trick is to have a plan. The tarpaulin is only to be the roof to your woodpile; you need to also think about keeping good airflow and fastening it down securely. Try to avoid any sagging depressions as these quickly fill with rainwater and form small stagnant pools – an eyesore and a breeding ground for mosquitoes. It is advisable to allow the tarpaulin roof to extend no more than halfway down the side of the wood stack, and it should be securely fastened with ties or weights.

Tarpaulins come in at least three qualities. If it is part of your plan to use a tarpaulin as a roof or sidewall to a woodpile, then I would strongly recommend buying a good one. There is a saying I like: "Buy cheap, buy dear", and this is never more true than when buying tarpaulins. In fact, there are so many maxims advising us against buying cheap things that it's a wonder people still do it. On this subject my absolute favourite is the wisdom offered by John Ruskin: "There is hardly anything in the world that some man can't make a little worse and sell a little cheaper, and the people who consider price only are this man's lawful prey." Determine the cover size needed for the job in mind, decide the most appropriate colour, and then buy a good one.

Solar kilns

The solar kiln is like a purpose-made greenhouse designed to use the sun's warmth to dry wood. The kiln must also vent the resulting warm, humid air or fungi will thrive in the damp atmosphere and start rotting the wood. Throughout most of my wood-burning life I have been nagged by the notion that I am missing out by not having a solar kiln. In preparing to write this book I was looking forward to the fact that, at last, I would finally have the opportunity to research solar kilns and identify the simplest and most efficient designs for small-scale log drying. Some people in Britain are using polytunnels as a form of solar kiln/woodshed. The logs are generally stacked on pallets within the polytunnel to season and not removed until they are required for the fire. If you have the space for a polytunnel woodshed this system may well suit you and there is no double handling. However, most solar kiln designs I have seen are something like a lean-to greenhouse, facing south to capture the maximum sun. All are designed to encourage airflow through the logs, in some cases using electric fans.

I read the summary report from an experiment in Alaska in which the moisture content of three groups of logs was monitored: in the open, in a conventional woodshed, and in a solar kiln. The weather recorded was similar to most British summers: occasional rain and roughly 20 days a month of cloud cover. The results surprised me. The wood left in the open had dried roughly 10% less than either the solar kiln or the traditional woodshed, but, amazingly, there was no difference in moisture content between the logs in the solar kiln and the woodshed. The researcher had taken great trouble not to overload the solar kiln, but nevertheless a proportion of the wood was showing mould and the early stages of rot. All of the kilned wood had been double handled for no advantage. I am not planning to build a solar kiln for my firewood.

CHAPTER SEVEN

SPLITTING LOGS

*"People love chopping wood.
In this activity one immediately sees results."*

—ALBERT EINSTEIN

There is nothing so wonderfully indicative of a wood fires lifestyle than a well-worn axe stuck in the edge of an old, battered chopping block. This simple and very rustic still life says quietly, but with absolute firmness: "There is skill here, a woodsman".

Like so many aspects of dealing with firewood, the key to good log splitting is to pace yourself and develop the necessary skills so that the whole exercise becomes an enjoyable workout, with the bonus of a heap of split logs when you've finished. I once read the maxim "bustle is not industry" on a Victorian ashtray, and over the years I have watched people, usually men, build up a huge sweat as they hack away at round logs. There are splinters of wood flying in all directions and the axe occasionally ricochets dangerously sideways, or more commonly gets jammed in the top of the log. There then follows five minutes of intolerable aggravation as the worker wrestles with the heavy and obstinate lump stuck fast to the head of the axe. As with most things in life, the skilful worker appears to be using very little effort and makes the whole process look really easy. So how do you achieve this very desirable level of skill? Let's take a look at the benefits, dangers, tricks, and know-how that turn log splitting from a chore into a pleasure.

WHY SPLIT THE LOGS?

The first and most obvious reason for splitting a round firewood log into smaller pieces is to reduce its size, making it easier to use on an open fire and easier to get into a wood-burning stove. A helpful subtlety with the open fire is that split logs are much less likely to roll out of the fire. When making a fire, either in an open hearth or in a wood stove, it is helpful to have a wide range of log sizes. The smaller pieces will help to get the fire started and, if you are trying to heat the room up quickly, again you will tend to use the smaller logs and split pieces. Another point to bear in mind, particularly with an open fire, is that the bark of several tree species has some fire resistance. You will get a better, quicker fire by always facing the clean split surface towards the heart of the fire; this will help the log to catch.

Once you have developed some accuracy with your splitting axe you can begin to choose the shape of the split logs. This is important for the fires you will make and is also helpful in the woodshed. There is no way I can build a reasonable log wall to edge my woodpile if the logs are all round (we look at the best way to do this in *Chapter 6*). Log-splitting patterns are shown later in this chapter. I make sure, whenever possible, that I split myself as many square and rectangular logs as I can. With these I can build a very stable edge to the wood stack.

Splitting to aid seasoning

Possibly the most important reason to split logs is all about drying: the seasoning process. I've said that the bark of some tree species is mildly fireproof, but in almost all cases it is very waterproof. I once bought a pile of beech logs that had been felled two years previously, since when they had lain on a sandy forest track in the dry climate of East Anglia. I naturally expected the wood to be fairly dry after two years and was astonished to find that in fact many of the logs were pretty much as wet as the day they were felled. When I looked more closely, it seemed that the cut surfaces on the ends of the logs had dried rapidly and shrunk, effectively sealing the ends of each log. The logs were roughly 4–5m (13–16ft) long and on many of them I found that the bark was undamaged and intact. So the log ends had sealed, the bark had prevented any further moisture loss, and unknowingly I had bought a heap of essentially green firewood! If you find that your logs are green and you cannot wait while they season normally as you need to burn them soon, split a proportion smaller than usual and these smaller pieces will then dry more quickly.

In all cases splitting the logs will speed up the drying process, but in some tree species this is more important than in others (for a summary of the drying properties of some common tree species, see *Chapter 5*). The cell structure of wood varies from species to species and the ability of water to pass radially through the log is very limited in some trees, whereas those that are diffuse porous are able to dry more readily as the water is able to pass through tiny holes in the cell wall radially out of the wood, as well as lineally along the rising sap tubes that run lengthways through the log. Birch, a favourite firewood of mine, is a special case in that the bark is astonishingly waterproof and the timber very susceptible to rot. So, if you leave birch logs "in the round" (i.e. not cut up) for any length of time, the chances are that you will lose some, or even all, of your firewood to fungal decay; notwithstanding that, birch is a diffuse porous species.

WHEN TO SPLIT LOGS

In my experience, the logs of all tree species split more easily when they are green. The logs don't have to be freshly felled – just split them before they have dried out. With species that split easily, like ash, you will hardly notice any difference between wet and dry logs when splitting, but remember that "efficiency is intelligent laziness" and if you want to use the least effort to split the most logs, split them green. I have heard it said that beech splits more easily when dry; if so, the difference is subtle and I have never noticed it. If you are unlucky enough to have a high proportion of knotty wood or crotches from where a tree has forked, then most certainly work them green. Once they are dry you may not be able to split them with an axe or maul at all, and will have to resort to wedges or cutting them up with a saw. I find the most difficult part of the tree to split is the stump. To give the tree extra strength at the fulcrum where the stem becomes the root system, the wood will often become cross-grained, wavy, and an absolute devil to split – again, you're off to the best start if you split these pieces as quickly as possible, saving the toughest one as your next chopping block.

Other factors to consider

You should split conifer logs when they are seasoned. One reason for doing this is that these woods when fresh are very resinous. I find that after even a relatively short period splitting freshly felled conifer logs my axe, hands, and clothes are sticky with resin, which then collects all available dust and dirt to form hard-to-clean black patches. For those people living in areas with really cold winters it is worth noting that logs split well when frozen – except the really knotty ones which are best left to thaw out again. However, if you find yourself splitting frozen logs on a cold day be aware that your axe handle may soon be covered in a thin coating of ice. My final thought on the question of when to split is for the person actually doing the work. If you are only cutting enough wood to heat your own home pace yourself and enjoy it. My grandfather was very fond of the saying: "A pint of beer a day is very good for you, but seven pints on a Saturday night does you no good at all!" I believe the same goes when splitting firewood: the occasional half-hour of steady splitting through the late winter and spring is much better for you than trying to split a whole winter's worth in one weekend.

WHERE TO SPLIT LOGS

There are three things to think about when considering where to split your logs: mess, damage, and safety. The area around the chopping block will become trampled and muddy, littered with fragments of bark and tiny slivers of wood. My cottage is small, with a tiny front garden and limited access to the rear, so I have to chop some of my wood on my precious front lawn. I move the chopping block from time to time to prevent the ground becoming compacted, and have to pick up every tiny piece of bark and wood splinter as they get stuck in the blades of my cylinder lawnmower. Where possible, I do my log splitting in the woodland where the firewood is being produced – but this is only because I do not have a good area for splitting logs at home.

Potential damage from splitting logs

The steady repeated blows from a splitting axe send quite an impact shock into the ground below and can cause damage. Before living in my current cottage I lived in a small farmhouse with a concrete drive. Believing the concrete to be strong, I used it as my chopping area until I noticed that I had put large cracks all across the drive. I will not now chop wood above any drains or other services that I believe may be close to the surface. Another more obvious and immediate cause of damage would be if

somebody were to chop wood in the proximity of a greenhouse or French doors; chunks of wood really do fly sometimes. Some people have the space within a large wood store to do their splitting under cover – great for working on a rainy day. But I have watched a man working with my heart missing the occasional beat as his swinging axe narrowly missed roof beams, lights, and wiring. To work confidently and safely ensure that you have plenty of clear space around you and that your footing is nonslip and clear of clutter.

The safety of others

When splitting, make sure that you can see all around you and that nobody can walk up to you unnoticed. The nearest miss I ever had with an axe was while I was a student on the two-year forestry course leading to the Royal Forestry Society's "Woodsman Certificate".

The course was very practical and we spent days splitting firewood in a large, purpose-built barn. I was absorbed in the steady routine of splitting when I miss-hit a log; the four-pound axe ricocheted wildly sideways, but I kept a tight hold of the handle and the axe head swung in a violent arc to my left. I was shocked to suddenly find a man standing there watching me, just out of sight to my left, and the ricocheting axe brushed the front of his wellington boots, leaving a faint line mark on them both.

Had the man been half a step further forward I think he would have had two badly broken legs. Imagine the horror if it were a child – nobody should come near you when you are working. Work where you can see around you and where you are visible and always be alert when chopping. Clear the area around your feet too so that you don't trip up.

Who should split logs?

Log splitting is often seen as heavy work, best left to barrel-chested, burly men. This is, of course, rubbish. All three of my children could split firewood from about the age of ten. I spent time with my two boys and girl teaching them to swing the axe safely and let the whipping force of the axe head do the work. They learned to give each log an inspection and put all the knotty, difficult ones aside for me. I did have to mend the occasional axe handle as they over-hit the log and damaged the shaft.

HOW TO SPLIT LOGS

Knowing how to split firewood should give you a great deal of pleasure once you have mastered the basic technique and can use your splitting axe or maul with confidence. Fundamentally, there are three ways to split a log: with an axe; a sledgehammer and wedges; or using a motor-driven mechanical splitter. The sledgehammer and wedges are only usually used on the most difficult logs and few homes have a motor-driven mechanical splitter. Most people burning firewood will use an axe to split some or all of their wood. Even if you have your logs delivered ready-split from a firewood merchant, there may well be a few of the larger logs that you wish to make smaller, so you may still need to do a little axe work. For the reasons above we'll start by looking at splitting logs with some form of axe.

Just before we do, it is interesting to note that in the past difficult logs could be split in a much more exciting way. In my late teens, when I had just started working in forestry, I was taught the basics by three old woodsmen. Each was nearing retirement and had worked the woodlands along the River Dart in Devon all their lives. Once, when we were trying to split a really tough old log with wedges, they said how much easier it would have been in the old days. Apparently a once-common technique was to drill a hole, or a line of holes, in a really awkward log and then fill each hole with gunpowder. A slow-burning fuse was carefully inserted and the holes sealed with clay. The woodsman then lit the fuse, stood back, and blew the log apart! Nowadays I imagine the authorities would frown on this technique and we are obliged to split our difficult logs with wedges or a hydraulic ram.

Staying slightly off the subject for a little longer, you may be interested in another historical note. There is a good deal of excellent firewood in a large tree stump, but because of the possibility of pieces of metal, dirt, and stones embedded in the wood, we tend to leave them to rot in the ground; the embedded debris would wreck a chainsaw and the decaying stump is excellent for wildlife. But once, to use a phrase from A.D. Webster's *Practical Forestry*, a "skilled estate hand capable of using explosives" could go to their local hardware store and buy sticks of dynamite. These large stumps could be lifted with the skilful use of the dynamite and then probably split using gunpowder, as just described. I've had a long career in woodland work, but sadly it did not include either of these techniques.

The axe
....

What a wonderful tool the axe is, perhaps the only tool used in earnest throughout history – from the Stone Age to the present day. It's amazing to think that the axe has evolved through stone, copper, bronze, iron, and steel, and is still used by people of all cultures, all over the world. So, it is a tool well worth mastering, if only to become part of this incredible history. There are many different types and weights of axe each designed to do a different job. For most of my life I have used a very blunt four-pound axe to split all of the easy logs and a six-pound splitting maul for the difficult ones. The important thing here is that you are comfortable with it; choose an axe of a weight that you feel you could use for up to an hour without straining or becoming overtired. In fact, I would strongly advise anyone to stop splitting wood as soon as you are tired: you are much more likely to lose accuracy and work less safely.

Don't be tempted to buy some fancy-shaped axe – these are almost certainly designed for tree felling or snedding. Pioneers used double-headed axes, where one blade was sharp for cutting and the other blunt for splitting and severing roots. But this is a specialist historical design – stay simple and safe. I mentioned that my four-pound axe is blunt; this is intentional and important in log splitting. Tree felling is different – before working with a felling axe a woodsman will hone the cutting edge until it is razor-sharp. The splitting axe, however, cannot be too blunt.

When chopping wood, the axe will occasionally get stuck in the log and getting it out again is frustrating and a complete waste of time and effort. So, when you strike a log, ideally you want the axe to split it or bounce back a couple of inches; a very blunt axe will do this much better than a sharp one. The splitting maul is really just a very, very blunt axe – a heavy wedge on the end of a handle.

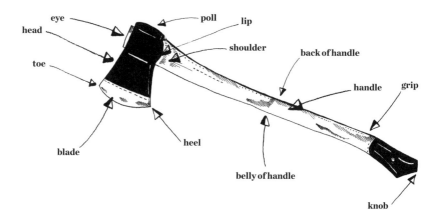

eye   poll   lip

head

shoulder

back of handle

toe

handle   grip

blade

heel

belly of handle

knob

Parts of the axe

*A splitting axe would
normally be roughly 2kg
(around 4/5lb). Always
remember, a felling
axe is razor sharp, but
a splitting axe must be
very blunt – or it will keep
getting stuck in the logs.*

A couple of years ago, a young friend persuaded me to try a
modern Finnish splitting axe. These are designed solely to do
this job and I am delighted with it. My old four-pound axe tends
to stay in the tool shed now. A good axe or maul should serve you
for many years if looked after; keep it dry, check the handle for
splits, and watch the axe head doesn't become loose. Don't use
the back of the axe as a hammer, as this widens the eye and the
handle will come loose. I've said that frozen wood splits well,
but just a note on a very cold axe: the old woodsmen would try
to warm the axe head a little before use in really cold weather,
as old axes consisted of a folded laminate with a high carbon
steel centre. A very cold axe can be brittle, causing the cutting
edge to chip, although this is more of an issue if it is a felling axe.

When using a maul it is still worth having an axe to hand to
finish the split in a log that is very fibrous. If the split is being
held together by thin splinters, an axe will finish the job more
easily than the maul.

While considering axes it is worth noting that there are now
modern innovations for splitting logs that work in the same way
as a slide hammer. The splitting wedge head is placed on the log
where it is to be split and then a weight, or part of the handle, is
used to deliver the blows that will split the log. I have used a slide
hammer and found the noise disagreeable, but this is an option
for those not comfortable using a traditional axe or maul.

Your clothing

....

A few moments considering the best clothing to wear when splitting wood is worthwhile – partly to make sure you are comfortable while working, and also of course for safety. Let's start at the ground and work up. Your boots should have a good grip and steel toe caps. I prefer a fairly long boot as this also provides some protection above the ankle; I have heard of people wearing shin pads while chopping wood and that seems perfectly reasonable to me. Bits of wood will occasionally jump back and hit you in the leg and, while not life-threatening, they certainly hurt! Your trousers, shirt, jersey, and/or jacket should be old, comfortable, and loose fitting. You need great freedom of movement to swing the axe and for all the subsequent stooping to pick up the split logs. Bear in mind also that being warm will help to prevent muscle strain.

There is some debate on the question of gloves, which I do not understand. If you want to develop real accuracy and skill with an axe or maul, you cannot do it through gloved hands: gloves are for mavericks and novices. There is also a very serious safety question here. When I was being taught by experienced craftsmen, gloves were forbidden when using any woodland hand-cutting tool. Tools are far more likely to fly out of your hands if you are wearing gloves, especially if they are wet from rain or dew. Staying with the hands for a moment, I also always take my wristwatch off when splitting logs. It is not that my watch is anything special, I just feel that an hour or so of jarring vibration would not do it any good at all.

The eyes are one of the most important parts of the body, and yet I've known very few people to use eye protection while splitting logs. Eye protection is recommended on many training courses and it's certainly something you should consider very carefully. One caveat, which I will also mention later, is that if you plan on using iron/steel wedges then I would strongly recommend eye protection. When hitting these wedges with a sledgehammer (as I said earlier, don't use the back of an axe), beware fragments of shrapnel-like iron that can fly, and these could blind you if your eyes are not protected.

The chopping block

A little time spent carefully choosing your chopping block will pay handsomely, especially in places like the UK where many of our logs are knotty and difficult. While it may appear to be little more than something to stand your logs upon, ready to be split, it is a key part of the splitting process. The log is being split between two things: the axe and the chopping block. For the axe to work completely effectively, the chopping block has to be a stable equal and opposite force. It should not be too small, nor wobble, its height should be correct for you, and, of course, it must be tough and not split after a few dozen blows.

I mentioned earlier that the lowest part of the tree's trunk, the buttressing, is always pretty tough and whenever possible my chopping block is taken from this part of the tree – an oak stump if possible. I choose oak mostly because of the weight; it makes an excellent, heavy chopping block that is not apt to move while I'm working. The spread of the buttressing also makes the bottom of the chopping block wider than the top: a great aid to the block's stability. I like to work with a chopping block that has a top diameter of at least 30cm (12in).

The height of the block is a very personal thing and will depend to some extent on how tall you are and also where you would like the split logs to fall. However, there are some logical advantages to keeping your chopping block low – around 15cm (6in) below your knee height. This gives the block a low centre of gravity, thereby increasing its stability, and will also give the arc of your swinging axe a few more degrees just before hitting the log, which will increase the axe's speed and give you more splitting impact.

A physics rule, which I think is true but don't understand, is that if you double the speed, you quadruple the impact. A low block also means that you are not lifting each log to be split quite as high – important if you have a lot to split. And, finally, if logs or fragments fly they are less likely to hit your shins with a low block.

But I did say the chopping block's height is a personal decision and a friend of mine has his chopping block slightly higher than his wheelbarrow. This allows a proportion of the logs he splits to fall directly into the barrow and thus saves him a little work. You will soon develop little tricks and preferences and these will become part of your style; if you are happy with them, and as long as they are safe, it doesn't matter that other people may be working completely differently.

One of my little tricks is that the upper surface of my chopping block has a slight slope of around 10°. When firewood logs are cross-cut using a chainsaw the cut is often at a slight angle and not perfectly perpendicular to the length of the tree. If your chopping block is perfectly level you may then have trouble balancing a log with a sloping cut surface on it – particularly with narrow logs. I find that having a slightly sloping chopping block allows me to stand almost all of my logs upright.

If a great many of your firewood logs have been cut at an angle, unlikely but not impossible, there is another technique I've heard of that you could try: fit a small car tyre around the chopping block to stop unstable logs falling off while splitting.

Positioning the chopping block is largely a matter of common sense. In all aspects of woodland work care is taken not to unnecessarily handle wood twice. The block should be positioned so that the split logs fall in a way that is helpful to you and requires the least handling. The working area should be level, not slippery, and away from any precious plants, fragile garden furniture, or buildings. If you are working under cover, rehearse the swing of the axe to be absolutely sure that it will not catch on any part of the roof or other fittings while you are working. Also ensure that the area to the left and right of you is clear, so that in the event of the axe ricocheting sideways nothing, and nobody, is damaged.

One final note on chopping blocks is to maybe have two. I sometimes place two logs on two chopping blocks, one in front of the other, and split the log nearest to me first, I then have a second log to swing at and split without letting go of the axe. In this way I halve the time spent letting go and recovering the axe, and put two logs up while I'm crouched with two hands free setting logs to be split. I feel it saves me a little time.

Accuracy with the axe

Within reason, it is true that when splitting wood any split is better than none. The cut surface will speed up the seasoning process and splitting gives you a wider range of firewood log sizes; this range helps when first lighting and then maintaining your fire. However, I am convinced that having adopted a wood fires lifestyle, most people will find that developing some skill with a splitting axe is highly desirable in terms of efficiency and safety, and it certainly makes the whole thing a lot more enjoyable. When I was training we would practise by putting a mark on a large, tough old log and then, using fairly gentle swings, would try and hit this target. To be effective when splitting you pretty much need to be able to hit the same spot most of the time. If you miss the target mark by more than 12mm (0.5in) either way then this is not good enough. I remember practising this as a student to achieve good accuracy, along with the other trainees. In time, this became a competition in which we would jam a matchstick into the middle of the log and then, using our felling axes (which, as I mentioned earlier, were razor-sharp), we would try and split the matchstick – amazingly, we often could. This level of finesse is not necessary for everyday axe work, merely being the sport of a small group of highly competitive youths, but it does show how accurate you can be once you focus. To be accurate and safe, it is also a good idea to check from time to time that the axe head has not become loose and that the axe handle is sound and not developing any fractures that might cause it to break while you are working.

There is a trick to picking up a splitting axe with very little effort. If you are right-handed, hold the end of the axe handle with the right hand, while the axe head rests on the ground, blade facing to the right. Grip the handle and move your right hand up and leftwards, then sharply back across to the right. With practice the axe will move up into the left hand with one smooth movement and very little effort.

Learn to use only the force that is needed to split a log. Accuracy and speed are key to efficiency, but hitting the logs too hard is a tiring waste of effort. Until your accuracy is good, it is worth placing the log to be split in the centre of the chopping block; the block will then stop the axe from dangerously following through. I always got my children to smoothly bend

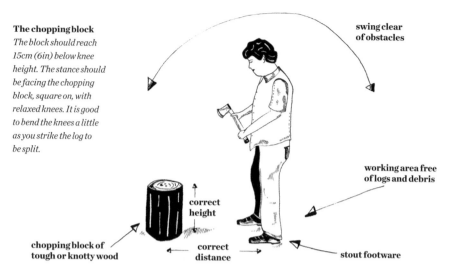

The chopping block
The block should reach 15cm (6in) below knee height. The stance should be facing the chopping block, square on, with relaxed knees. It is good to bend the knees a little as you strike the log to be split.

swing clear of obstacles

working area free of logs and debris

chopping block of tough or knotty wood

correct height

correct distance

stout footware

their knees as the axe was about to hit the log, to prevent the axe from coming back at them if they were to miss-hit, which does happen when you are learning. Also, work with the longest handle you are comfortable with – this helps with speed and limits the chance of it coming back at you. One final thought on good technique: stand facing the log you want to split and, as far as possible, avoid working with a twisted back.

Log shapes and other specific techniques

Once you have a reasonable level of axe skill and can strike a log pretty much where you want, then you can begin to cut your firewood into your preferred shapes and also deal with the more tough, difficult logs. I cut my firewood logs into pie-shaped pieces, with some squares, rectangles, and slabbing as shown in *figs. iv* and *v* on page 148. As you pick up each log to be split and carefully place it on the chopping block, you should run your eye over it, looking for any knots, crotches, or other problems. This is good technique as it can save a lot of splitting time. *Figs. i, ii* and *iii* (overleaf) show some typical difficult logs and how to approach splitting them.

As I noted earlier while discussing the best time to split wood, all logs split more easily when they are still green, and if you live somewhere with really severe winters you may find that wood also splits well when frozen. A third general point would be not

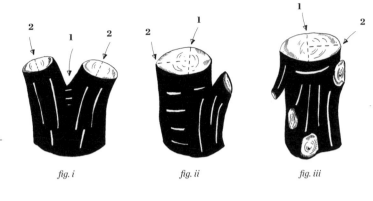

Typically difficult logs
Fig. i shows you how to first split down the centre of twin trunks, then each now separate log. Fig. ii shows the split down the centre to cut the log in half, avoiding the knot. Fig. iii shows a log upside-down, a good technique with knotty logs. You make the first cut furthest from the knot.

fig. i   fig. ii   fig. iii

to be stubborn: some logs just don't want to be split. If you have a really awkward log that, after a reasonable effort, still shows no sign of splitting, then put it to one side to be split with a hammer and wedges or cut up with a chainsaw later. Although just before you give up, try turning the log over – some logs for no obvious reason just split better from one end.

Having looked at some general wood-splitting points, let's now look at specifics. A log with a large diameter can be "peeled" by taking slabbing off the edges, or a weak line can be created across the centre and then the log cleft along this line. If a knot is found, that is the point at which a branch emerged from the tree's main stem; place the log "upside-down" on the chopping block, thus cutting "upwards" into the tree. If the way the branch was growing is not clear, then place the knotty end of the log on the chopping block, leaving the clearer wood to split first. Similarly, a "crotch", where there was a fork in the tree, is also best dealt with by being placed upside-down on the chopping block. However, sometimes – when using a splitting maul – I have found that hitting repeatedly in the absolute "V" of the crotch splits the log quickly. The one exception to this is when splitting an actual stump; in this case it is better to strike the log with it aligned the right way up. If your logs are coming from a really knotty tree, and you have any say in the matter, have the logs cut into lengths shorter than usual as this will make splitting them easier. Lastly, a tip on picking up logs one-handed that

> **Having enthused about the scent of the different tree species' wood smoke, I would just note how lovely newly split logs can smell. For me, oak is rich, warm, and yeasty, while birch is fresh and sweet like apples.**

may save you a little pain. If you have long fingernails, pick the log up so that your fingers wrap around the circumference, not by holding the log end on. Particularly when the log is damp, the bark may be slightly slippery and if you hold the log by its end it may suddenly slip, bending your fingernails backwards. Alternatively, simply use both hands to pick the logs up.

Stuck logs

From time to time the axe will get stuck in a log, although this rarely happens with a maul. There are five basic ways to free the axe.

1. Try tapping the end of the axe handle sharply downwards with the heel of your palm – obviously not hard enough to damage the handle, or yourself. If the axe is only just stuck, this action will usually free it.
2. The next stage is where the log is not too big to lift and tapping the axe did not free it. Lift the axe with the stuck log wedged on the end, turn the whole thing over and drop it on the chopping block. If you feel that the log will split, hit the axe on the block so that the weight of the log drives it further onto the axe, hopefully splitting it.
3. If you just want to get the axe out of the log, hit the edge of the stuck log on the block so that the weight of the axe frees it. But take care, as with methods 2 and 3 the log will often suddenly free itself and fall off the axe.
4. With a large log, or if the axe is stuck fast, first try tapping the axe head gently with a lump hammer. Tap each way in line with the axe blade – this will usually loosen the axe.
5. The last resort, and to need this you have somehow got your axe really stuck, is to take a hammer and wedge and set the wedge in the log in line with the stuck axe. By carefully hammering in the wedge you should open the split the axe has started, or at least take enough pressure off the axe to be able to now tap the axe out.

After freeing an axe always look at the axe head and handle to make sure there is no damage and the head has not loosened.

Log splitting patterns

It is easier to cut larger diameter logs into a mixture of squares, rectangles, and pie-shaped pieces. Smaller diameters can contain more pie shapes (fig. iv). Large logs contain more rectangles (fig. v), which can be useful when stacking.

fig. iv

fig. v

Splitting kindling

The secret to making good kindling is to choose your logs very carefully. When I am putting logs in the woodshed I am always on the lookout for ones with straight grain and no knots whatsoever. These perfect logs I put to one side to be cut into kindling. In terms of tree species it is generally believed that conifers, such as pine, make the best kindling, but this very much depends on how you intend to light your fire – which we look at next in *Chapter 8*.

The simplest splitting pattern is to take the "mother log" and cleave off thin slabs – as if you were cutting roofing shingles. Then take each of these boards and, with the axe in one hand and the board in the other, chop off your kindling sticks. A kindling stick diameter of something like 2.5cm (1in) is perfect. Interestingly, there are four common ways that people split kindling – one of which is not safe.

I would say the most common way people chop kindling is to take a small hand axe, of say about one pound, and then, while holding the board in one hand, swing down with the axe to split off each kindling stick. This method works and is quick and easy as you are using such a light axe. However, the hand holding the log is obviously in some danger of being hit by the axe, and this danger increases as the board being chopped gets gradually smaller. A huge improvement to this technique is to use one of the kindling sticks to hold the board that is being cleft. This technique, rather ironically, is called the "sissy stick"! Another safe, but slow, method is to use a much heavier axe on a very short handle. The axe is positioned on the board at the point you wish to chop it and then the board and axe are lifted together and then dropped back down again. With the small axe the force needed to split off the kindling stick is generated by the speed of swing, while the much heavier axe allows this splitting power to be generated at a slower speed. Lastly, perhaps look at getting a froe and beetle as another safe way to split kindling.

ALTERNATIVES TO THE AXE

Using sledgehammer and wedges

In the same way as you may put aside your perfect logs to be cut into kindling, so you should put aside any really difficult logs (those full of knots, crotches, and forks) if you feel there is no chance of splitting them with an axe or maul. These logs are going to be a challenge! To break these difficult logs into usable split firewood you will need your axe, a sledgehammer, and two or three wedges. The best wedges are generally made of steel or forged iron, but you can also use wedges made of magnesium alloy or tough plastic. The wedges come either as a traditional "wedge" shape or a star-shaped variation. Firstly read each log, looking carefully for the knots, forks, and cross grain, and choose where your first cut should be using the same splitting rules as described under Log Shapes on page 145. I would always recommend wearing safety glasses when splitting with a hammer and wedges, and do not let anyone hold the wedge in place for you.

These difficult logs are often large and I tend not to put them on the chopping block, preferring to work them on the ground. Having chosen where I think the first split should be, I tap the log hard enough with the axe to make a slot in which to position the first wedge. This wedge is then driven in using the sledgehammer until the log begins to crack. As soon as the fracture line forms in the log, set the second wedge in this crack a little way from the first, then tap the second wedge in too. You now have two wedges in the same fracture and should hammer

each in turn until the log splits. If the log is huge, you could use a third wedge in the same way. If the log turns out to be impossible, and you get the wedges stuck, free them by knocking them out sideways with the hammer and then put this obstinate log aside to be cut up with a chainsaw. If you prefer not to try and work out how best to split a difficult log, the star-shaped wedges might be best for you. As you hammer one of these in, it puts pressure in all directions and should split the log along the line of least resistance. A little maintenance is required with wedges as over time they may develop burring where they are struck by the hammer. This is where fragments may fly from and so it is good practice to grind off any burr that is developing.

Powered wood-splitters

Once, powered log-splitters were the preserve of professional woodsmen or those with a workshop and the know-how to make one, but not any more. If a powered log-splitter appeals to you and you think that this may be the way you want to go, then there is a very wide variety to choose from. They are stocked by some large garden centres and suppliers are found via the internet.

Most have fundamentally the same design. The log to be split sits between a solid strong point and a strong moving point, and a hydraulic ram slowly exerts a very high pressure between these two, causing the log to split. Basically, a splitting maul-type wedge or axe-type blade is driven into the log by hydraulic pressure. There are three power options: with the most simple, a hand pump creates the hydraulic pressure, while the more common models are ones in which an electric- or petrol-driven motor drives the hydraulic ram. If a tractor is available there are many models that use the tractor's PTO (power take off) to drive them. Some splitters are designed to work vertically, but most seem to work horizontally. They are easier to split logs with than a maul, but they are far from effortless. You still have to pick up each log, place it on the splitter, and manoeuvre the splitter into position. Also, the rams on splitters work slowly – a good thing for safety, but if your wood splits easily it may be quicker to use an axe. I'm told working is more efficient if you have two people: one to place the logs and one to operate the controls.

*

CHAPTER EIGHT

THE WOOD STOVE & OPEN HEARTH

"With Leaves and Barks she feeds her
Infant Fire: It smoaks; and then with
trembling Breath she blows, 'Till in a
chearful Blaze the Flames arose."

—"THE STORY OF BAUCIS AND PHILEMON", JOHN DRYDEN
METAMORPHOSES VIII, OVID

**We have looked at all the preparations for a life with wood fires –
the trees, the wood, purchasing, storing, seasoning, and splitting the
logs – and now I can hardly suppress my own trembling breath as at
last we come to the very essence of the subject: making the fire itself.**

I love fire. I can't quite put my finger on why I'm so content, to my very soul, being with this radiant companion throughout the long winter. I said in my introduction to *Chapter 1* that having a real fire is something like having a much-loved pet – you will make it a bed somewhere central in your home and then carefully tend and feed it, and clean up after it. Our very language seems to casually assume and attribute life to fires. On finding a cold hearth, the fire is said to have died or gone out, as if fire were a benign spirit that has now left the wood and charcoal. In creating your perfect fire you will, with utmost care, nurture and feed the glowing infant, this incandescent sprite, until, like a parent who sees their offspring grow to robust early adulthood, you lean back and relax as "in a cheerful blaze your flames arise".

We most certainly live in interesting times, wood fire-wise. During the million years or so that we have been able to create fire at will, and to have domestic fires, I feel there have been three significant evolutionary steps. To have an open fire within our cave or home, where the smoke finds its own way out as best it can, accounts for 99.99% of our time with fires. Over the last 0.01% of time, chimneys became widely used, and then the enclosed stove was developed. Both were huge improvements in wood burning. Now, in the early 21st century, a fourth evolutionary step is taking place: burning cleanly with much greater efficiency in heat recovery. For me, living through this next step and reaping the rewards is just wonderful.

In this chapter I want to look at the skills and tricks needed to make and tend a perfect wood fire, using the everyday knowledge of our forebears, together with new research and design knowledge, which are so important as an enlightened renewables age dawns. Wood-burning stoves have become dominant where firewood is an important part of a home's heating plan, and most stoves are also beautiful, whereas the traditional open-hearth fires now tend to be reserved for special occasions and celebrations. I will look at both, their similarities and critical differences, and later in the chapter I will cover subjects common to both, such as firelighters and tinder, lighting a fire, and the physics of heat.

But, just before we move on, I would offer you the early memories of Laurie Lee from his autobiographical book *Cider with Rosie*. His moving prose is second to none in reaffirming and reminding us of the singular importance that the home fire has had for so many, for so long: "... most of Mother's attention was fixed on the grate, whose fire must never go out ... The state of our fire became as important to us as it must have been to a primitive tribe. When it sulked and sank we were filled with dismay; when it blazed all was well with the world; but if – God save us – it went out altogether, then we were clutched by primeval chills. Then it seemed the very sun had died, that winter had come for ever, that the wolves of the wilderness were gathering near, and that there was no more hope to look for."

THE WOOD-BURNING STOVE

In the new cottage I am building I expect to burn less than half of the logs I currently burn, and to be warmer. I will cook on a wood-burning range that is around 80% efficient and approved for use in smoke-controlled areas. A small modern wood stove with a clear ceramic glass door will grace my living room – also approved and about 70% efficient. This is me becoming modern, soon to be living in a well-insulated home and still enjoying all the pleasures of my wood fires, but burning far less wood, far more cleanly. I might not even need my winter vest!

The wood-burning stove's secret is that it allows the airflow to the fire to be controlled and a thorough "secondary" burn to take place, meaning that all of the inflammable vapours and carbon (smoke) rise from the logs to burn above the fire itself. As around half of the heat energy in a log is in these vapours, this is really important. This is called "clean burning", as there is little or no smoke leaving the chimney. When choosing a stove, look for a model with a secondary air supply and, if you want to see the flames, a high-quality glass door. All models have a primary air supply.

The logs

The wood-burning stove is altogether easier and more forgiving than the open fire. While the logs still need to be well seasoned (≤ 20% moisture content), they can be good-, intermediate-, or poor-quality firewood – they will all burn well in the wood stove. The tree species also matters less because logs prone to throwing sparks can be used. Furthermore, a burning round log cannot roll out if the door is closed. I find having logs of mixed lengths helps to create a fire with air spaces and turbulence. Just because a stove will take, say, 38cm (15in) logs doesn't mean that this is the right length for the stove.

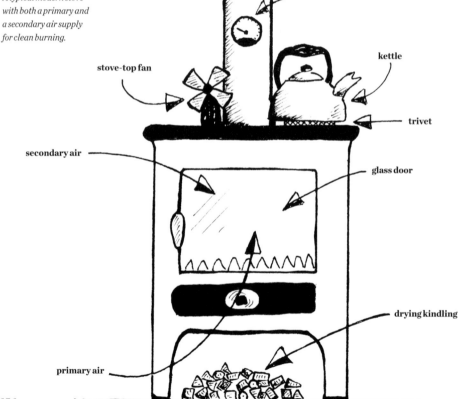

Modern stove
A typical modern stove with both a primary and a secondary air supply for clean burning.

thermometer

stove-top fan

kettle

secondary air

trivet

glass door

drying kindling

primary air

The air

The stove will take the primary and secondary air in balance with how you set the intake vents. The manufacturer's guidance is the best place to start, until you have learned to fine-tune your own stove. Many people find that keeping the door slightly open when the fire is first lit provides the abundance of air the early fire needs. The stove must be able to draw air in easily. You may need to have air bricks, an underfloor draft, or a direct air feed into the stove if your home is airtight and may not readily allow adequate air for combustion. If it will not draw when first lit, try pre-warming the chimney flue with burning paper to get the warm air rising and a draft started. In an airtight house there may be back pressure if an extractor fan is running. My grandmother would often open a window slightly when first lighting her fire – she seemed to understand this need for "easy air" instinctively.

The fire

The wood stove is a hot box that radiates heat into your room. The flames and embers both contribute to this and here the most important thing is to see a lively rolling flame burning all the smoke. So, manage your fire, through the air intakes, to maintain this hot, steady burn. Neither create a furnace-like fire, so hot that it will cause damage, nor shut down the air to the point where a dull red gossamer flame struggles to keep going. If you can see smoke rising out of the chimney, or your glass door blackens frequently, you have not got it right yet. When adding logs to the fire, open the door just a little at first, to help to prevent a surge of heat, sparks, or smoke coming into the room. A good friend of mine enjoys hearing his stove fire crackle and leaves the door open. This effectively negates all the benefits of a stove and makes it a normal open fire – I think he does it to annoy me when I visit. The only downside I can see to having a stove is that there is a sheet of glass between you and your fire and you can't hear it, but that's a small price to pay.

Much as I love my wood stove, I am ruthless in wanting efficiency from it – I want the maximum heat at the minimum cost to my woodshed. In the early stage of the fire, say for the first half hour or so that it is burning (phase one), I want the stove, the firebricks within it, and the iron flue pipe above to heat up and become evenly hot. I then feel that the fire is comfortable and bedded in. I am generous with the logs in this early stage, as thriftiness will produce a little runt of a fire that will need more care and nurturing later than I want to give. Once I feel that the fire has achieved my bedded-in point, I change tack completely. From then on (phase two) I feed it a basic ration of, usually, two logs when the previous ones have almost burnt to embers. As the hours pass on a long winter's day, I seem to use very little wood, but still keep my stove hot. I use my larger logs during the long phase two. When everything is really hot within the stove, I squeeze in the biggest logs I can and relax, knowing that they will burn for well over an hour. The last time my "door open" friend visited me, I wanted to emphasize the efficiency of a hot, closed-door stove and refuelled my fire with one large, bone-dry, elm log. He seemed genuinely surprised when it lasted for three and a quarter hours – the log sat on a deep bed of hot embers and always had enough air to burn cleanly.

By all means show prudence in your use of wood, but not meanness – you must try to maintain the critical heat at which the stove's fire can cleanly consume large logs, and not smoulder them away. Be mindful not to emulate poor Bob Cratchit in Charles Dickens' *A Christmas Carol*: "Scrooge had a very small fire, but the clerk's fire was so very much smaller that it looked like one coal."

During phase one, while getting my stove fire going, I will open the door a little, or open the vents to double what I then set them at for the prolonged phase two burning. I manage the fire through the amount of fuel I give it – not the air. I never really restrict the air to my stove as I want to be absolutely sure that all the chimney-damaging acidic tar and creosote given off by the logs is thoroughly burned in the stove. I know many people who are content, even proud, to make their logs really last by severely restricting the stove's air, but this is a bad practice and I would urge you not to do it.

....

Avoid also emulating the fire scene from the film *Withnail and I*, in which, sitting by a small, damp fire in a windswept farmhouse, Withnail observes: "We may as well sit round a cigarette. This is ridiculous. We'll be found dead in here next spring."

....

Overnight burning with a wood stove

With the wood-burning stove I would first suggest that you read the manufacturer's instructions – if they recommend overnight burning, take careful note of how they suggest you do it. In my own home I make no effort to keep my stove lit overnight, not even in the depths of winter. It is absolutely no problem to relight it in the morning. A man whose opinion I trust once told me that the best technique with a wood stove is to build up a large quantity of embers and then, when absolutely sure that all the tarry vapours have been burned out of the wood, close the air inlets a little. Allow just a trickle of air, enough to keep the ember bed glowing through much of the night.

The overnight burning hazard is subtle and the danger here is that partial combustion results in the smouldering logs releasing a steady quantity of gaseous wood tar and creosote, which then condenses on the lining of the chimney. The dangers are multiple: firstly, this liquid is highly acidic and will eat away at the chimney's mortar and metalwork, but there is also the possibility of a build-up of soot and wood tar causing a chimney fire. The enormous heat generated by a chimney fire can crack the lining and walls of the chimney and in some cases lead to a house fire. I was once shown the inside of a huge chimney where the whole inner surface glistened with glassy wood tar and there were actual tar stalactites hanging down – the previous house owners had been proud at being able to keep their stove going overnight! I believe that the practice carries a real risk and should only be undertaken with great care.

There is another possible way to keep the fire going through the night, which, to my knowledge, is only used in the harshest climates. While staying with a family in the Yukon Territory in north-west Canada, I was keen to learn how they fought the bitter cold through the long winter. They told me that when the temperature dropped below something like -30°C (-22°F), they set their alarm clocks to wake them every four hours, so that they could load more firewood into their boiler. They did not dare allow the intense, stabbing cold to penetrate their home and freeze everything. This must be a hard and relentless routine during a period of prolonged cold, but it is an alternative. So, the other way to keep your fire going through the night is to just wake up regularly and feed it normally.

Cooking on a wood stove

I was at first concerned that my home would constantly smell of whatever I was cooking, as the smells could not be drawn up the chimney. I couldn't see a way around that, but then, after frying a hearty breakfast on my stove, I found that I actually rather liked my cottage smelling of sausages. So, when making a meal on my stove, I have accepted that my home will occasionally be scented with lovely cooking smells. I also keep a tough kettle on a trivet on my stove. The "free" hot water is appreciated and it does add a bit of moisture to the air, which can become very dry when the stove is going.

When I first started wood-stove cooking I limited myself to a frying pan, but over time I learned how to use different-sized trivets to give me the more gentle heat needed for slow cooking. I now make a cracking beef stew on special occasions. I roast whole shallots and cherry tomatoes in a frying pan with the lid on, next sautéing some carrots and celery in butter. I chop a small swede into little cubes and then sear strips of seasoned beef in the wood fire's flames. Everything then goes into a large cast-iron casserole dish with a rich beef stock and a spoonful of hummus, along with the secret ingredient – three succulent prunes. Cooked for several hours, I would say, immodestly, that this recipe makes one of the best stews I have ever eaten.

I am careful never to place a pot or pan directly on the surface of my iron stove for fear of cracking the top, and I always use trivets when cooking. But be warned: in the event of a spillage, it is not possible to turn the stove off and clean up the mess. So I take great care when cooking, for while I relish the savoury, mouth-watering smells of good food, I don't

I haven't yet cracked the problem of baking or roasting meat joints on my wood fires. I do have a Dutch oven (a large, cast-iron cooking pot with a heavy lid) that I use for roasting and baking while camping, but it is too large for use on my stove indoors and deflects too much smoke into the room when I've tried to use it with the open fire.

want the reek of it burning. Actually, and as a quick aside, I do sometimes deliberately place a coffee bean, scrap of birch bark, or a few grains of frankincense or amber onto my hot stove – there is no danger of these tiny offerings damaging the stove and my room is quickly filled with their wonderful scent.

THE OPEN HEARTH

When I move into my new home I will certainly miss my open fire. My last three houses had big open hearths and I could build a fire to the size that suited the occasion. One Christmas the yule log I had carefully selected was so big and burned so fiercely that my family had to leave the room – I was disappointed and felt I'd scored something of an own goal. I had chosen a huge resinous Scots pine log, cut from where the tree had snapped in a gale some years back, and had expected cheering compliments on my fire skills. Instead I was scowled at.

Perfect fireplace

1. Seasoned logs
2. Fire-resistant glove
3. Underfloor draft
4. Firedog
5. Wood ash fire base
6. Cast-iron fireback
7. Chimney swept
8. Companion set
9. Kindling
10. Bellows
11. Front clear for spark guard

There is much joy to be had in the open hearth, and your fire will need you to have real skill in selecting and placing your logs. With care your fire can burn fairly cleanly, but it will only ever achieve a low efficiency – something like 15–20%. In general, if the fire is to be lit on special occasions as a treat, that is fine – but obviously not in a smoke-controlled area. While the key to good stove fires is the management of the air drafts, the secret to good open fires is the management of the glowing embers.

There is a huge variation in open hearths, from vast manorial inglenooks to the tiny cast-iron Victorian coal grates. The fire bed itself may be flat or raised, on an iron grate, or held by "firedogs". Ever since burning coal became commonplace in homes, there has been confusion over the correct management of the ashes. While the coal fire's ashes and clinker are a nuisance and should be cleaned out daily, I am a great believer in the benefits of having your wood fire sitting on a bed of its own insulating and reflective ash. I would say resist cleaning out the ashes until they become a nuisance, and don't use a high grate where the fire flames well but the all-important embers fall below the fire and cannot radiate their heat. I was amused to read in an old copy of the *Times* a letter to the editor on the subject of open-hearth wood fires, in which the correspondent described a visit to a friend who had a large house in southern England. He noted that: "By the hearths of the living rooms were pinned notices calling down the most fearful imprecations on any housemaid who should remove the least particle of ash. At the time I wondered at the violence of the language; now I understand." The correspondent had moved into an old house and during two bitterly cold winters he made an intensive study of his wood fires to ascertain what was most important in maintaining them. He decided that there are two fundamental rules to having good wood fires, and a decent bed of ash is the first of them.

Every open fire is pretty much unique, as the draft feeding it will be affected by a number of things, such as how the air is drawn into the room, the nature, height, and design of the chimney, any surrounding trees and other buildings, and, finally, the weather. You must learn all the idiosyncrasies of your fire before you can be confident of lighting an excellent fire time after time. But, again, this is something to enjoy – this fire is your friend and to get along together you must learn its moods. I'm suddenly reminded that mastering wood fires is similar to falconry: you have tamed something very wild and must learn its ways to succeed – the fire will be very slow to learn yours. And if fires have been my falcons, I would add that I carry several small scars where in moments of carelessness I've been bitten!

The logs
....

Put aside any large logs of the best dense firewood you have as the back log – species such as elm, oak, beech, hornbeam, field maple, or robinia. This will act as a reflector and help to radiate heat into the room. You will need some small, very dry logs of low- to medium-density species to get the fire going quickly with the least smoke – birch, alder, ash, hazel, and sycamore work well. If you choose conifer logs to start the fire, that's fine, but be ready for some sparks to be thrown. The logs needed to keep the fire going normally can reasonably be of any species that does not frequently throw sparks. I only use split or crooked logs to help prevent them rolling out – split logs also catch fire faster.

The air
....

There is very little control of the airflow with an open fire, but a few things are important. If the initial draft is poor, try warming the chimney before you light the fire – a torch made of rolled-up newspaper should work. Or hold an open sheet of newspaper across the mouth of the upper part of the fireplace to send what little draft there is across the kindling, and so help the fire to develop. In an airtight house, opening a window slightly may help. But this fire is going to need to draw in a lot of air as it burns, and if it can't do that you have a problem. Not only will the fire often be difficult to get going, but there is more chance of smoke and fumes coming back into the room – and with them the deadly gas carbon monoxide. If there is an underfloor draft duct built in, make sure the grille covers are fully open. I have read that the base of an open fire should be around 25cm (10in) above floor level, to allow air to flow smoothly in through the fire and up the chimney. Some fireplaces have an iron "damper" to seal off the chimney flue. If yours does, then keep it fully open when first lighting the fire. The damper is then used to control the chimney draft and can be closed off during summer or when the fire is not in use.

The fire

....

To provide really meaningful heating, the open fire should have as many glowing embers radiating warmth into the room as you can reasonably manage. This is where the open fire is so different from the stove and requires more care and skill. The enclosed stove fire makes the entire surface hot, which then radiates heat and is difficult to block, while the embers in an open fire have a much smaller surface area and their radiated heat can easily be shielded from the room. I rather like natural history writer and poet Marnie Reed Crowell's observation that: "To keep the fire burning brightly there's one easy rule: keep the two logs together, near enough to keep each other warm and far enough apart – about a finger's breadth – for breathing room." She goes on to complete this excellent advice: "Good fire, good marriage, same rule." In fact, the *Times* correspondent I mentioned earlier gave much the same advice with his second cardinal rule of open wood fires: "The logs must lie close together – kissing." The critical point to remember is that when tending the open fire you are really managing the bed of embers, while when tending the wood stove you are primarily managing the air draft.

People love to see lots of flames in their fire, which is good for clean burning, but the secret to an effective, really warming open fire is your management of the ember bed. Time and time again, I see people feed their fire by placing a large, cold log right in the middle of it. Apparently, this is exactly where the logs are burning low and where the fresh fuel is needed, but it feels like somebody has turned the fire off – suddenly there is no sun-like warmth on your hands, legs, and face. The way I introduce new logs to my open fire is to move the half-burned ones together into the centre with my poker, then I place the fresh logs on the left and right of the fire, cut surface facing inwards, bark surface facing outwards. This helps to achieve two important things.

Firstly, I make sure that the new logs do not prevent the embers from continuing to radiate heat into my room – the sun-like fire does not go behind a log-like cloud. Secondly, the logs from the wood basket are at room temperature and are, therefore, relatively cold. Laying the fresh logs on the edge of your fire gives them the chance to pre-warm and dry a little before they actually become the fire itself.

I have assumed so far that your fire is for heat, but of course there may be times when you are primarily seeking to create flames, and that most evocative ambience – firelight. The light of naked flame is sublime and enriching and gives a quality to a room and its furnishings that harsh electric light can seldom achieve. It does the same for people: our friends and lovers in firelight are given a new beauty, their skin toned as if in the last few minutes of summer's setting sun. Reflecting quietly on this, I feel that perhaps the shadows are equally important – ever moving, emphasizing some features while hiding others – and what delicious shadows we make. To create pleasing firelight, your wood basket should be well loaded with fully dry, small-diameter logs of the denser and more brightly burning species, such as beech, ash, hawthorn, oak, and the harder maples. If you have some, put aside apple or cherry wood, box, or walnut for such evenings.

Also learn not to waste your wood by using too much – don't make a fire bigger than you actually need. "It takes time to show people the mistake of piling up logs. It is easy to waste three times the amount of wood needed," writes W. Robinson in his lovely little book *My Wood Fires and Their Story*. He goes on to say: "The old people who made the fires and had no choice as to the fuel learned how to make the most of their wood." Like them, we should learn how to make the most of our wood.

Mark Twain's friend Charles Dudley Warner found that, "To poke a wood fire is more solid enjoyment than almost anything else in the world." You too may enjoy giving the fire a good poke from time to time and seeing a shower of bright sparks swirl upwards, but the wood fire does not like it. Coal fires need to have air passing right through the burning embers, which is why coal is burned on a raised grate and the fire needs poking regularly as the pieces begin to fuse together. Wood fires, on the other hand, are more refined and not to be idly prodded and poked.

I have used a trick to amuse and maybe puzzle occasional visitors, tapping a couple of copper nails into a log. This will create a patch of gossamer blue-green fire among the normal flames – but don't overdo it or you will simply make the fire seem haunted!

Overnight burning on the open hearth

Many people take great pride in their ability to keep their fire "in" overnight. Before 1826, when John Walker invented the safety match, it was an important skill with open fires. I believe the historical importance of keeping a fire in was not to have warmth through the night, as there is very little warmth from a smouldering log or embers buried under a heap of ashes. They kept their fires in to allow easy rekindling from the embers in the morning – making a fire from sparks or friction is much harder work. However, while the open fire may be kept in overnight, there are reasons to be careful as, unwatched, it may escape the hearth during the night. If you choose to keep it in, be clear on why – we have matches now.

Cooking on an open fire

I find there is only limited scope for cooking over my open fire. If I use anything more than one kettle or pan, my room fills with smoke. This is probably why one only ever sees a large, rounded iron pot hanging above the open fire in illustrations of traditional kitchen life. The outer surface of my pans blacken and I have found that the old trick of first coating them with soap doesn't work! Also, the heat from an open fire is more erratic than the stove – it is well suited to boiling and frying, but not much else. To grill over an open fire you need to build up a large heap of embers and then let a thin covering of white ash form before you place your meat or fish over it, otherwise the outside will burn. A lively open fire does of course offer you the chance to make the most beautiful toast or muffins.

FIRELIGHTERS & TINDER

I enjoy the frugal challenge of maintaining a small bin of home-made fire lighting materials and thus limiting my use of manufactured firelighters. My bin could easily be mistaken for a rubbish basket. It does not enhance the aesthetics of my fireside, but I cannot resist a magpie-like urge to collect the many and varied little scraps of clean inflammable material that will help to start my fire. I do have a pack of bought firelighters in the cupboard, as a backup. On the rare occasions when I use one, I cut the stick in half as that is plenty to get my fire started. There is a weak environmental argument against using firelighters and, of course, they are an unnecessary cost, but in truth I just collect the scraps because I enjoy it. I am careful to avoid plastics – why risk the acrid stink of burning plastic in your room when you can so easily have the scrumptious smell of birch bark, orange peel, or pine cones?

I've mentioned that dried orange peel, or any of the satsuma, tangerine, or clementine group of citrus fruits, makes a good firelighter. I find it best to peel the fruit carefully into a starfish shape for drying on my hearth. Birch bark also makes an excellent firelighter, but it has the habit of curling up tightly as it starts to burn. So I interlock my pieces of birch bark, like the links of a paper chain, and find that this keeps them open and helps them to burn more efficiently.

Paper is the most obvious material to start a fire, but during my lifetime the nature of paper, newspaper in particular, has changed and it is nowhere near as good for fire lighting as it once was. My understanding is that modern newspaper is coated to allow for colour printing and the sharp definition of

Firelighters

A dried orange peel "star" (fig. i). Birch bark curls, linked to limit curling (fig. ii).

fig. i

fig. ii

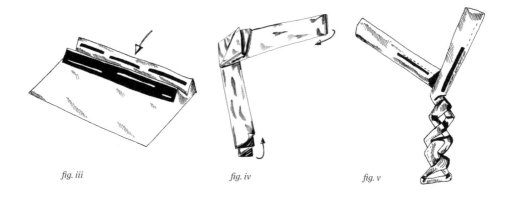

fig. iii   fig. iv   fig. v

Paper concertina

A sheet of paper is repeatedly folded to form a flat stick (fig. iii). The flat stick is now plaited by folding at 90° to form a "concertina" (figs. iv and v).

Hollow hoop

Roll a sheet of paper into a thin tube (fig. vi). Bend the roll into a hoop and twist the ends together to secure the shape (fig. vii).

text. I believe the coating is largely china clay – a product of the decomposed rock feldspar – which is unhelpful in fire lighting. Brown wrapping paper and cardboard contain a much higher proportion of wood fibre than newspaper and are more useful in starting your fire.

It was always my custom to simply screw up a few sheets of paper into rough balls, but my mother-in-law showed me how she was taught as a girl to make paper sticks, which then extend out like little concertinas. She laid a sheet of newspaper flat and then folded it again and again, each time carefully pressing the folds, until it was a long flat stick no more than 2.5cm (1in) wide. She then folded it 90° at the centre and, as if making a two-strand plait, she folded each end alternately across the other, until she had made the concertina. She would use a few of these together to light her fire. My grandfather had a different system and would carefully roll his sheet of newspaper into a thin tube with an airspace down the centre. The tube was then gently bent

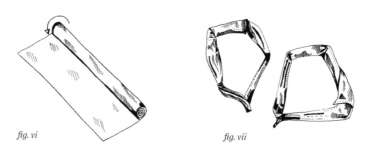

fig. vi   fig. vii

into a circle with the ends twisted together to keep its shape. In the fireplace they looked vaguely like a little heap of white horseshoes. Another shape that works is to roll a paper tube and then scrunch the ends, like little Christmas crackers.

We looked at the kindling itself in *Chapter 4*, but I would add that if you have oak logs and the bark is coming free, then pull off chunks of it to make excellent kindling – the bark of most other trees is really rather thin to be considered as kindling.

An unsavoury habit I have that improves my oak bark's fire lighting is to pour old bacon fat onto it when making my breakfast. I picked this tip up from reading *The Second Meadow*, Archie Hill's book about his time living rough in a woodland.

Striking a match

With the fire laid and matches in hand, kneel or crouch close to the fire. Visualize exactly where the burning match is to go. Take out a single match and hold it close to the phosphorus head between thumb and first two fingers – to reduce the chance of the match breaking. Place the match head against the striking surface and, keeping the match at a shallow angle, strike it in one smooth movement. Once the match is alight and the candle-like flame is steady, bring the match to the horizontal so that it does not burn too quickly. Cup your other hand to provide a windshield as you move the flame to the fire-lighting material. You have up to 15 seconds to use the match before the flame is too close to your fingers to be comfortable. If a fire is prepared properly, it should only take one match to light it.

LIGHTING THE FIRE

It is hard to change the habits of a lifetime – especially as we get older, which I am now. But there has been a revolution in the simple and timeless act of lighting a wood fire, and over the last year I have been practising the new technique in both my stove and on the open hearth.

Research has shown that lighting a fire using the traditional "bottom-up" approach is when it is most likely to create smoke. This is because the weak first flames are low in the fire, underneath the kindling and some small logs. The early heat is enough to create smoke, but there is no flame above the fire yet to "secondary burn" the smoke, and it slowly drifts up and out of the chimney. This is perhaps OK for a lone farmhouse, but in an area where many homes have wood fires this is now unacceptable. The answer, thankfully, is simple – just adopt the new "top-down" technique to light a fire. By inverting the way a fire is laid and lighting it at the top, the early flames create very little smoke and more readily burn any that

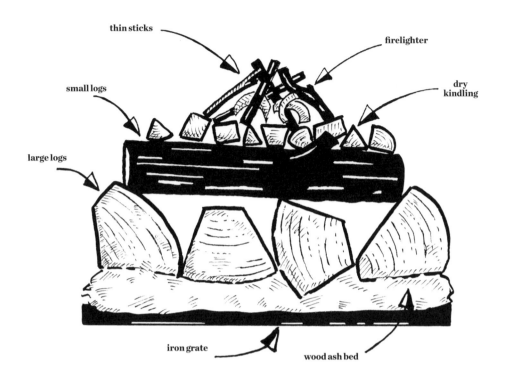

thin sticks

firelighter

small logs

dry kindling

large logs

iron grate

wood ash bed

Top-down

Building a "top-down" fire creates a much cleaner initial burn.

does form. I was sceptical of this at first and saw it as a novelty fashion idea. But it does work, and as we strive for cleaner air, this improved technique is a godsend.

To lay a top-down fire, first place three or four medium-sized logs, slightly apart, on the bed of the previous fire's ashes – split surface upwards. Then at right angles to these lay a row of smaller logs. It is best if this row is of less-dense, quick-burning species and about 5cm (2in) in diameter. Again at right angles, now build a platform of very dry kindling sticks on top of the second row of small logs. Finally, using your chosen fire-lighting material and a few very thin sticks, build a small pyramid fire on top of the kindling platform. It all looks a bit like a short fat candle when first lit, but then the kindling quickly catches and the fire burns downwards – with almost no smoke. I am sure that this will soon become the classic fire-starting technique for home fires. The clever thing is not to load fuel above the early fire and cause smoke. In time, variations to this basic technique will develop and these will be fine too. In a small stove it may only be possible to lay one row of logs at the base, before building the kindling platform, and this is fine as by the time all the wood is burning the stove will be hot.

I usually light my fires with a safety match – I leave friction and sparks to bushcraft enthusiasts. Although the standard small matches are cheap and easily bought, you may find that you are happy to pay a little more for the stronger, longer matches. These really do make home fire lighting easier as they allow you to position your first flame exactly where you need it. You may like to go one stage further and buy a butane candle lighter – the hotter flame is easier to direct. But, in all honesty, if you have good kindling material, ordinary matches will be fine.

Lighting the fire from embers

In the many centuries before matches became commonplace, a home fire was normally lit by using the carefully saved embers from the previous day's fire. Lighting a fire with sparks or friction takes effort and great skill, particularly in cold and damp climates. For this reason, most early peoples probably kept their fires constantly alight and used the old embers whenever a new fire was needed. When a fire had to be moved, the hot embers or pieces of smouldering fungi were carefully carried nestled in leaves, or perhaps in a hollow plant stalk – Prometheus fashion.

I enjoy kindling a fire from embers, it enables me to practise my fire skills and hone this timeless technique. In the morning it is fun to gently scrape away the ashes and see the hot coals still glowing – it feels good, like unwrapping a present. This was once the habit of almost all people, of almost all cultures, all over the world. As with the woman in Ovid's poem at the beginning of this chapter, when lighting a fire using embers, you will always need to blow on them to get it started. Bring the hot embers together into a small heap, and blow on them gently if they are only partially alight so that they all glow brightly. I find it is best to add any dead pieces of charcoal to the embers at first to get the biggest and hottest heap of coals possible – this is a smoky technique and starting with more embers speeds up getting a flame. Then carefully place pieces of thin kindling, mixed with birch bark, orange peel, small twigs, torn cardboard, wood splinters, and whatever else you have saved for fire lighting, over three sides of these red-hot embers, leaving the side facing you open. Crouch low, so as to blow horizontally through the open side onto the embers in long, slow breaths. Do not give hard, short puffs, nor breathe so hard that you become light-headed. If you have long hair, take care to pull it back as your face will be close to the fire while you are doing this. Those people with bellows or a blow poker may find the whole thing more comfortable, but again I would stress long, gentle blows. Once the first flame appears you can relax a little and add more pieces of kindling until the whole thing is again a cheerfully blazing fire.

The open fire was the staple heating and cooking system for almost all homes for centuries across Europe and North America, and keeping the open fire in overnight was absolutely standard practice. There are recorded examples of people keeping a fire alight for their entire lifetime. A correspondent to the *Times* in 1929 noted that, "The predecessor of the Forest Inn at Hexworthy [on Dartmoor in south-western England] was said to have possessed a fire which for over 100 years had never been allowed to go out." It is probable that really old fires were once quite common and that the kitchens of many houses would keep their winter fire going on into the summer for cooking, baking bread, and warming water.

Small bundles of dry sticks called "faggots" were traditionally kept in the kitchen to make rekindling the fire as quick and easy as possible. The faggots were also used when any room was to be heated quickly. I read of one house that used two faggots of hazel and one of ash to warm the "withdrawing room", to be ready for guests leaving the dining room.

I had expected that ancient fire-lighting folklore would have some reverence for old fires, but curiously this tradition seems to have been quite rare. The more widespread view was that a new fire was very special – often referred to as the "need" fire. This tradition reinforces the probability that people would keep their fires alight for as long as possible. But it was not simply a matter of efficiency: old fires were considered tired and weak. A new fire, kindled fresh from the labour of sparks or friction, was thought to have a desirable wildness, with youth and vitality.

CONDUCTION, CONVECTION, & RADIATION

Heat energy is not straightforward and I used to be baffled by how, on the one hand, heat cannot cross a vacuum – hence a Thermos flask working. But then, on the other hand, the sun's heat readily crosses a vacuum of 150 million kilometres (93 million miles) to earth and warms us. An understanding of the three ways in which heat moves will help to build effective fires.

Heat moves through three basic processes – conduction, convection, and radiation – and these all work together in providing available heat. With an open fire there is only a limited opportunity for the heat to conduct and it will do little more than creep into the cast-iron fire back and the bricks that make up the hearth and chimney. All the metal of a wood stove and flue pipe will become hot by conduction. The next process, convection, is to do with heat moving within a flow, in this case of air. The heated air close to the fire is less dense than the surrounding cooler air and rises, causing a circulation of air within a room. This gentle convection is helpful unless you have a very high ceiling, in which case much of your precious heat may end up a few feet above you.

The third process, radiation, is by far the most important. When the fire is burning well on the hearth, or in the stove, heat is transferred as electromagnetic radiation across the room – exactly the same process that brings heat from the sun. This explains why the warmth from a glowing fire feels like sunshine. Conduction and convection play their part, but the heat radiating from the fire is what really matters. That is why, with an open fire, I stressed the importance of maintaining a good bed of glowing embers to radiate warmth and not suddenly putting a log right in the middle of the fire. The wood stove is heated by conduction and then transfers heat to your room through the moving warm air, convection, and, most importantly, by the hot stove itself radiating heat in all directions.

CHAPTER NINE

*

FLAME, SMOKE, EMBERS & ASH

"O flames that glowed!
O hearts that yearned! ... "

—*THE FIRE OF DRIFT-WOOD,*
HENRY WADSWORTH LONGFELLOW

While I am inordinately fond of trees, axe-work, and sheds full of beautifully seasoned logs, it is the flames, smoke, and embers that are the fire. These three, this timeless tripartite, are what all the earlier preparatory work was about.

The ash is obviously something else entirely, but although it is the result of the fire, and not the fire itself, the ash too has a role and is worth a closer look. There has been a great deal written recently about the pivotal role of fire in mankind's evolution and how we are physically and socially adapted to fire; it is then no wonder that we are so at one with it – a curious and unique symbiosis. I have even heard it said that fire exhibits most of the proofs of life, the seven biological processes common to all living things: growth, movement, respiration, reproduction, excretion, sensitivity, and nutrition. Now that's something to consider quietly one evening while watching your living fire!

The saying goes that fire is a good servant but a bad master. I believe that the fire you make should be much more than simply a good servant. When about to light a fire, consider first the purpose for lighting it and what the flame/smoke/ember balance should be to meet this purpose. Any wood fire, however badly made, is likely to produce flames, smoke, and embers, but the skilful fire-maker will manage the fire to produce exactly what is needed. So rather than just setting fire to a heap of wood, sitting back, and accepting how it burns, you can learn to make each new fire burn how you want it to and produce the flame, smoke, or embers you need. That is the fire-craft, the fire husbandry; like a shepherd moving sheep, you must master the knowledge of how to nudge, cajole, and move your fire's three-element balance to what best meets the need you created it for. As with most skills, the true craftsman makes it look easy and effortless – the fire just burns beautifully, seemingly all by itself. But there is much more to flame, smoke, and embers than first meets the eye – when analyzed, they turn out to be incredibly complex.

In previous chapters we have concentrated on the fire's fuel: wood. Now let us look carefully at how and why you might manage a fire to create and maximize bright flames or glowing embers, and consider the wood fire's smoke.

FLAMES

It is the flames (from the Latin *flamma* and old Anglo-French *flaume*) that many people seem to enjoy most about their fire. The aesthetic beauty of firelight is undeniable; pubs and hotels attract us by boldly advertising a "blazing wood fire". The very light itself seems cheerful: the red-orange glow, the ever-changing pattern of dancing flames, and the animated shadows they cast. The gentle light of flames is calm and forgiving, only illuminating the features that we want to be emphasized; like a scene from an Old Masters oil painting, it leaves in shadow and half-lit gloom that which is not in the forefront of the scene.

I once lived without electricity for 11 months when the supply to my cottage was cut off – a difficult time! I burned a lot of candles and did an awful lot of thinking in the flickering glow of my wood-burning stove. Reading, even by the light of a dozen candles, was really difficult. So, even with flame, perhaps you can have too much of a good thing; try to keep it special. Happily, I never tired of the flames themselves. In normal daily life we should only ever need to use the light of flame when we really want to – such as for special dinners or self-indulgent baths – and therefore flames and firelight should never lose their timeless, magical luxury.

What is a flame?

This is one of those questions that can leave the enquirer feeling slightly stupid or guilty, as the answer is surely pretty obvious. A flame is just, well, a flame! We are all familiar with them – whether as a dancing crowd vibrantly alive in the midst of a bonfire or the solitary beauty of a single candle flame. Flame (fire) was seen as one of the four elements by the ancient world and, in a way, their categorization of it as an element belies its complexity. Beware of simple explanations; the processes taking place even in a single candle flame are extraordinarily complicated – deliciously so for the chemist or physicist, and almost bafflingly so for the woodsman.

It is probably helpful initially to look at the process that results in a flame. Firstly, there needs to be a fuel – wood in our case – and available oxygen. These two happily coexist in every woodland and forest all over the world without regularly bursting into flame, and so it is obvious that a third thing is needed: heat. The heat does two things: at around 120–150°C (248–302°F) the wood cellulose begins to decompose and emit a volatile gas mixture, essentially smoke. Then, as the temperature rises to around 250–300°C (482–572°F), this vapour reaches its ignition temperature and catches fire. The magic of fire is that, unlike most chemical reactions, this one is self-perpetuating. Once the original ignition heat source is removed, the reaction is producing enough of its own heat to vaporize more fuel and thereby keeps the reaction going. This reaction is rapid as well as incandescent (luminous and hot). Rusting is an oxidization process too, but a slow one and not remotely as interesting.

The preceding paragraph is a sketch of the process, but now let's look more closely at the actual flame itself, this body of incandescent gas. I have said that with sufficient heat the decomposing wood emits smoke, and this smoke is primarily composed of hydrocarbons. So we have hydrogen and carbon locked in cellulose and lignin organic molecules, and we have oxygen within the surrounding air. The combustion reaction combines the hydrogen with the oxygen to form water (H_2O) and the carbon with the oxygen to form carbon dioxide (CO_2). It is the movement of atoms from the hydrocarbon molecules and these same atoms' recombination to form the new molecules of water and carbon dioxide that releases energy in the form of heat and visible light. The colour of a flame changes with rising temperature, as the molecules move to a more excited state and emit energy

Flame colour variation

Variation in colour within a flame is caused by uneven temperature. The lower part is generally seen to be blue and is the hottest, while the cooler, upper part of a flame will burn yellow/orange. The blue flame seen above burning charcoal and in the hottest part of the fire is the gas carbon monoxide taking on another atom of oxygen to become carbon dioxide. A yellow flame is basically burning soot particles. The colour of flames in a fire will also change in the presence of other chemicals or metals; as mentioned previously, a few copper nails tapped into your logs will create patches of mesmerizing, mysterious, blue-green flames. Sodium colours a flame yellowish and the bright yellow-orange flames of driftwood fires are to a large extent coloured by the sodium in the sea salt that has dried into the wood.

at a higher frequency. So a flame is the distinct reaction zone in which the vaporized fuel is oxidizing and the visible flame is exactly where this molecular realignment is emitting heat and visible (as well as infra-red) light. An interesting aside is that the shape of a flame is governed by gravity; in conditions with a very low gravitational force (microgravity) a flame is spherical. There are thought to be literally hundreds of subtle variations to this basic chemical reaction taking place as a wood fire burns, due to the enormous variation in the initial fuel hydrocarbons and the fact that within a burning fire much of the oxidization is incomplete, but sometimes only temporarily so. Your fire is an incredible and almost immeasurably complicated flow of chemical reactions. Watch the next candle flame you see with a renewed sense of childlike wonder – it will not be misplaced.

The best woods for flame

If you are building a fire primarily for flame and light then the first rule would be to select the driest wood possible. Avoid rotten wood, which will only ever smoulder, producing a good deal of smoke, but no flame. I would say the next most important thing is to resist the temptation to try and burn large-diameter wood. Choose logs of say 7.5cm (3in) diameter or less; if you have larger logs you want to use, split them into smaller pieces. If, after being mindful of the above (dry, not rotten, and small-diameter), you still have a choice of wood to make your bright flaming fire, I would pick any of the following tree species: beech, any maple, ash, hawthorn, birch, and pine. But really, it is the state of the wood (dryness and size) more than the species that is important for making flames.

Building a fire for flame

We talk about building and maintaining a home fire in *Chapter 8*, but while on the subject of flames it is worth just noting the most important factors in helping a fire to make flames. You already know that the wood you are using should be dry, small in diameter, and not rotten. The best fire structure to build is a pyramid, as this will give the fire height and allow a good supply of air to the burning sticks. Have lots of prepared wood ready, as the fire will burn through the supply very quickly. Importantly, feed the fire a stick at a time where gaps are forming, be careful not to choke it by putting on too much fresh wood. Keep the sticks 2.5–5cm (1–2in) apart, as this is close enough for them to reflect heat, but far enough apart for plenty of oxygen to get to them.

SMOKE

Wood smoke (from the Old English word *smoca* or *smeocan*) generally gets a bad press, often referred to as "choking", "stinging", or "acrid". The problem is always with having too much smoke. As when wearing perfume or flavouring a meal, a touch is lovely but an overload isn't. In small quantities, the various woods have the most amazing and delicate perfumes, as we saw in *Chapter 2*. After lighting the wood-burning stove in my sitting room I'll often take out a small, smouldering log and hold it in the air for two or three seconds, just long enough to scent the room, and then return it to the fire.

Wood smoke is also very beautiful to watch: the swirling ethereal patterns, slow-moving, mesmerizing – I find it curiously comforting. I'm not alone in this, and when Ulysses was trapped on an island by the unhelpfully persistent goddess, Calypso, apparently he could "think of nothing but how he may once more see the smoke of his own chimneys".

I remember experiencing a similar feeling once when I was doing some geological work on the west coast of Ireland – not of being trapped, but of the gently evocative images and yearnings triggered solely by the scent of smoke. It had been a long day, and I was cold and wet as I walked into the main street of Westport in County Mayo; it was dusk and soft rain was still falling. The street was empty and although lights were already on in the houses and pubs, it was curiously quiet. I wanted to get back to my lodgings and out of my sodden clothes. Then, the smoke touched me and brushed passed like a friend – perhaps a light gust of wind or an eddy over a building had carried it down into the street. I stopped to savour it, the rich, almost fruity scent of burning peat (always

called "turf" by those that burn it). This brief encounter with the merest scrap of peat smoke caused an emotional impact that was much more intense than my simply enjoying a pleasant smell – it evoked images conjured by association. The scented wisps told of nearby warmth, dryness, food, and drink, but – perhaps most importantly as I had been working alone all day – the smoke told of people, company, and, this being Ireland, foretold music and laughter. But I needed rid of my wet clothing so shook off these tempting images and walked on; Ulysses had Calypso, I had wet trousers.

Anthropology may explain why most of us find catching the slightest whiff of wood smoke in the air so pleasing. It may at first seem curious that we enjoy this complicated and possibly carcinogenic cocktail of poly-aromatic hydrocarbons so much, but it is evocative of similar experiences to my own in Ireland – tired, wet, alone, I was delighted by a fire's smell, and to a returning hunter or gatherer it must have meant everything. To them, it was the first scent of wood smoke that was joyful: the herald of safety, food, family, and friends – perhaps it is this, an echo from our so very distant past, which makes a little wood smoke so unaccountably pleasing. There is a lovely little poem by Henry David Thoreau that takes delight in watching smoke; he clearly enjoyed seeing it rise from early morning fires and captured this with the lines: "Lark without song, and messenger of dawn / Circling above the hamlets as thy nest."

What is wood smoke?

I mentioned earlier that wood smoke has been getting a bad press over the last few years; I believe that as we learned smoking cigarettes was a major cause of illness, so research focused on the inhalation of smoke in general. But, just before analyzing what smoke actually is, let me stress again that we can produce little or no smoke by learning good burning techniques and, if possible, updating our open fire or stove to a modern clean-burning model. Most of the research data that I have read on wood smoke's chemical composition shows the enormous complexity of the smoke *when there has been incomplete combustion* – the result of insufficient air, or a "cool" fire caused by burning unseasoned wood, or both! This generalization is unhelpful. While true for wild fires and very bad home fires, it does not accurately represent the actual emissions from the chimney of a responsible owner with fire skills. A good fire will not produce smoke. Modern wood-burning stoves should be burning the smoke as part of the secondary burn, and a smoky open fire, whether indoors or outdoors, is just a nuisance. Remember, you waste a lot of heat energy if you allow your fire to smoke, so assuming you bought the wood, you are also wasting money.

What is it we actually see when we observe smoke from a wood fire? The visible smoke may contain any, or all, of the following three elements: water vapour, other liquid particulate matter, and solid particulate matter. All firewood will contain some moisture. Freshly felled timber has around 50% moisture content and this is taken down to 20%, or less, by the seasoning process. The moisture is driven off in the form of steam by the heat of the burning fire to then condense into visible water vapour as it cools. The more moisture there is in the material being burnt, the more of this "white smoke" is produced. As I say, this is in fact not smoke at all, but harmless water vapour – the same as condenses above a boiling kettle. The other liquid particulate matter is a very different story.

About half of the possible heating energy within a log of wood is in the form of wood-tar/creosote; like the water moisture within a log, this pyrogenous liquid is also vaporized by the heat of the fire. In a well-balanced fire this highly inflammable vapour will burn, but in a fire that is choked with too much fuel, or has an insufficient supply of air, the tars/creosote will first vaporize and then condense again later as they cool. These condensing droplets will either line the chimney if it is cool enough or condense in the air above the chimney as blue-grey smoke. If the smoke is very thick with these tar droplets it will appear yellow to brown in colour. The solid particulate matter in wood smoke comprises carbon and wood ash. Black smoke contains unburned carbon in the form of soot particles. The particles of ash are white and produce a visible white/grey element to wood smoke, but they may be so small as to be invisible to the naked eye and, therefore small enough to penetrate deep into the lungs.

In addition to the minute ash particles, another invisible element of wood smoke is the gases produced by combustion. A good fire with plenty of oxygen burning at a fairly high temperature will produce a very small amount of smoke, as almost all of the carbon is converted into carbon dioxide. A high-temperature fire may also produce oxides of nitrogen. Wood has a low sulphur content in comparison to coal and so the production of the gas sulphur dioxide is far less than that of a coal fire. However, a poor, smoky wood fire with incomplete combustion (oxidization) produces a wider range of chemicals and may produce the toxic gas carbon monoxide. Hydrogen may be produced instead of water. I remember once talking with charcoal burners, to whom incomplete combustion is an art form, and hearing them talk of the dangers of hydrogen forming in their sealed metal kilns. Finally, the mixture of pyrogenous vapour and gases formed by incomplete combustion is highly inflammable and can ignite on contact with an open flame; one cause of a chimney puffing smoke back into the stove.

The enormous complexity of wood smoke and the recent research into its possible consequences to human health seem at first to be a little concerning for the wood fire enthusiast, and put a bit of a dampener on the joy felt by working with wood fires. But I would stress again that you should not tolerate smoke leaking into a room from a wood-burning stove or an open fire, nor should you sit breathing in clouds of smoke from a campfire or garden bonfire. Take joy in your flames and pride in your smokeless fire. With a "no smoke" attitude to our sustainable wood-fired home heating, I believe we can confidently stay on the environmental moral high ground.

Smoke signals

Only the occasional involuntary castaway may now need to light a fire for the purpose of creating smoke to signal their distress. The use of smoke as a communication medium has all but passed into history, but it is worth recording just how important it once was. As a boy I never tired of the endless stream of "Cowboys and Indians" films shown on Sunday afternoons, in which the desert Indians were adept at communicating by the use of smoke.

These signals were sobering and portentous as they seemed to innately demonstrate the native people's superiority in the desert wilderness. The Australian Aborigines were able to make smoke in four colours – white, grey, black, and blue – and the smoke itself could be sent up as a solid column, long or short dashes, smoke rings, or spirals, enabling their communications through smoke to be remarkably complex.

The modern-day, small-scale charcoal burner still uses the smoke from his kilns to tell him how the burn is going. A crucial part of the charcoal maker's art is knowing when to stop the air supply to the burning wood. During the first stage of the burn the fire is building up heat and drying the packed wood by driving off the moisture; through this stage the smoke is white, the whiteness being a cloud of water vapour. Then, quite suddenly, the smoke will turn blue and this is the signal for the air supply to be closed off, preventing any further oxygen from getting to the burning wood and ruining the charcoal.

Smoke folklore

With fire having such a long history, one would expect a very rich and varied store of fire folklore – and there is. But there seems to be very little folklore concerned with wood smoke. However, one old Eastern European tradition I particularly like is of trying to stupefy witches as they fly hidden in storm clouds by sending up smoke made from the toxic concoction of laurel and wormwood. In a few parts of Europe, capnomancy (divination through the observation of smoke) was practised. If the wind took the smoke eastward, there was hope of a good harvest, but westward presaged a crop failure. Similarly in ancient Sweden, the direction taken by the smoke of the May Day bonfire was thought to be a guide to the coming summer – if blown to the south, it would be warm; to the north, it would be cold. Perhaps it was such ancient smoke fumigation/purification practices that led to the ritual use of incense smoke.

Smoke from recycled or reclaimed wood

I have already advised against burning any wood that does, or is likely to, contain preservatives, fungicides, pesticides, or paint. Any concerns over the smoke from untainted firewood logs must be greatly magnified when we are considering the smoke from recycled or reclaimed treated wood. These woods may contain a wide range of harmful metallic compounds, arsenic, or organo-chlorines, or possibly be heavily loaded with additional poly-aromatic hydrocarbons through treatment with tar or creosote. To me, the potential cocktails of undesirable toxins resulting from burning mixtures of treated wood suggests a simple rule – don't do it.

EMBERS

A fire's embers (from the Old English *Æmerge* "ember") just sit there glowing quietly, neither dancing nor changing colour: flame's bridesmaid or, perhaps more accurately, flame's old mother. There is nothing really mesmerizing about a slowly glowing ember and yet, as most fires are actually lit to produce heat, either to warm us or cook our food, the embers are in fact enormously important. Flames, for all their bewitching gossamer beauty, are actually shallow, transient, their entertainment drawing our attention from the real bedrock of a warming fire – the embers. As discussed in *Chapter 8*, the management of the embers is the secret of a good fire and the goal of good fire husbandry. It may well be that roughly 50% of a firewood log's potential heat energy is in the volatile vapours, but these vapours are really only effectively burned in a modern wood-burning stove or boiler, because these stoves are designed to give the heat and air needed to ensure an efficient secondary burn and release the heat energy from the rising inflammable gases and smoke. In the case of an open fire, whether in the hearth or as a campfire, you are unlikely to achieve a thorough secondary burn and so the greatest proportion of the heat available to you from a log will come from the fire's embers.

What are embers?

A glowing ember is as pleasingly simple as a flame is bewilderingly complicated. For a wood fire to burn, the volatiles are driven from the log by heat and combust above the fuel source in the form of diffuse flames. This process, after having removed the volatiles from the wood, leaves chunks of char, which is essentially carbon and a very small amount of wood ash material; this is the widely used charcoal. The charcoal, being almost entirely carbon, simply oxidizes during combustion. The carbon atoms bond with atoms of oxygen to form carbon dioxide, or, where there is incomplete combustion, the more dangerous gas carbon monoxide. Therefore a glowing ember is very different from a flame in that it is the fuel material itself that is burning right through and not a flammable vapour burning above the fuel source.

The best woods for embers

As is the case when wishing to produce flames from a fire, it is really important for the wood to be dry when you want embers. In this case, however, the diameter of the wood being burnt is not so important. In fact it is an old technique to use one large-diameter log as the back to a fire. As the fire burns, the surface of this large log chars and forms a layer of glowing embers. These then radiate heat straight out at those sitting in front of the fire. Once your fire is well alight I find that the denser woods, perhaps predictably, make the best embers. Favourites of mine would include hornbeam, beech, oak, elm, hawthorn, and fruitwoods.

It is worth singling out the elm for special mention. I feel that ash is overrated as a firewood, while the highly desirable elm is commonly and scurrilously dismissed as near useless. Sadly, after the terrible Dutch elm disease suffered by native elms across Europe and North America in the latter part of the 20th century, elm wood is not as common as it once was. However, nature does have a knack of fighting back and while the sight of a magnificent full-grown elm is now relatively rare, many hedgerows are still full of pole-stage young elms. But the disease still has the upper hand. Just as these promising young trees, which grow as suckers from the old roots, reach a few inches in diameter, the beetles that carry the disease start feeding on the twigs and cause infection, which kills the young elms. This is a great pity, but I have to say that every cloud has a silver lining. These young elms, their dead stems standing sun-bleached like driftwood scattered along the tide line of our hedgerows, do make extraordinarily good firewood – especially for embers. If you do manage to get hold of some of this elm, be pleased and use the logs in mixture – use them for what they are best at: making embers. As the embers are critical to a good cooking fire, building a fire for embers is covered in *Chapter 10*.

Folklore and uses of embers

The scraps of folklore that I am aware of mostly refer to the cold, "dead" embers, especially when these are from an important fire. However, living in a cottage with no central heating, an old use of embers that really appeals to me is for bed warming. You sometimes still see the old bed-warming pans in antique shops or on the walls of country pubs. Typically they have a long, wooden handle with a frying pan-like brass bowl and lid on the working end. On winter evenings the warming pan would be primed with a few glowing embers and then slipped between the sheets to take the chill off an icy bed. I understand that you had to keep the pan moving or the sheets could easily be scorched. The use of warming pans declined rapidly with the development of hot water bottles, which were easier and safer to use. My bedroom is generally fairly cold and for me one of life's most sublime moments is the first few blissful seconds after snuggling down into a cosy, pre-warmed bed. I confess to currently having an electric blanket, but with the price of electricity spiralling upwards, I might buy a warming pan and see how I get on with it.

Across Europe the May Day and midsummer bonfires were once extremely important and, in various ways, people felt they ensured the fecundity of crops, cattle, and local women. I'm always fascinated by the obscure and curious, and it's interesting to note that, with all the many ills and problems our ancestors had, a dead ember from the midsummer bonfire could be dipped in holy water and then carried as a protection against the unlikely ill of being struck by lightning. I have read many times that the rural peoples of Europe were enormously superstitious and so I guess it is reasonable that among their many beliefs there had to be a protection against lightning. More pragmatic and vague was the belief that if an old man put one of these special embers into his wooden shoe he would be protected from many evils.

WOOD ASH

Many people don't like the idea of cleaning out the fireplace or stove and cite this as one of the reasons why they are willing to forego the many pleasures that a real fire brings. However, unlike coal fires, wood fires leave very little ash or debris. Around 1% of a wood log ends up as ash (from the Old English *Æsce* "ash"), with the majority of that coming from the bark. The recorded percentage of ash content within the logs of different tree species varies from 0.43% to 1.82%; for practical purposes this is generally averaged to 1%. The ash content of bark is surprisingly high, up to 6%, and the ash content of pure stem wood is usually less than 0.25%. By comparison, the ash content of anthracite coal is typically 10% to 20% of its weight. What this means in practice is that the clinker-free ash from a wood fire is easily and much less frequently removed than is the case with a coal fire. But there are two other important points in favour of wood ash. Firstly, for a wood fire to burn well, in most cases it needs a decent bed of the soft white ash for the embers to rest in and radiate their heat – a wood fire is helped by a bed of its own ashes, a coal fire is not. Secondly, wood ash has a wide range of possible and interesting uses.

What is wood ash?

The composition of wood ash varies enormously depending on the type of wood being burnt, the temperature at which it was burnt, and whether or not there is any residual organic carbon (charcoal) left within the ashes. Typically the wood ash will contain between 25% and 45% calcium carbonate, up to 10% potash (potassium compounds), around 1% phosphate and the micro-nutrients iron, boron, manganese, copper, and zinc. There is no significant nitrogen within wood ash. The ash from hardwoods has more potassium and phosphate than that of softwoods, but less calcium and silica.

Uses of wood ash

The ash from wood fires is a real resource and over the years many different ways have been developed to make use of its properties. By far the most widespread use of wood ash today is in the garden. Wood ash offers two very real opportunities for the gardener – as a lime to reduce soil acidity and as a source of potassium and phosphate. It is often cited that wood ash can be used as a physical barrier in pest control and deters snails and slugs, especially where the soil is damp and it frequently rains. The consensus seems to be that wood ash works well in gently raising the pH of acidic soils and compost. Obviously plants that like a low pH will not appreciate being dosed with wood ash, and nor will potatoes as this may lead them to develop scab. On the other hand, ash will help with the problem of club root on your brassicas.

Historically, wood ash was most prized as a source of potash (potassium), but note that this is highly soluble and any ash that has been rained upon is likely to have had the potash leached out. If you intend to use wood ash in the garden it is important to keep it dry, as once wet it forms a slimy mass that is highly alkaline and difficult to handle. In the late 18th and early 19th century there was a massive industry burning North American trees to make ash for potash, which was then mostly shipped to Great Britain. This was a very useful source of income for the hard-pressed settlers and most of the trees burned were those felled while clearing the land for agriculture. This industry declined as other sources of potash were discovered. The abundant ash from coal fires has no value as a fertilizer.

After its use as a soil improver and fertilizer, two other important uses for wood ash were in soap making and as a glaze in pottery. I find it fascinating to imagine how mankind first learned that a cleaning agent could be made by mixing concentrated dissolved wood ash with animal fat. Perhaps it was simply that over millennia people noticed that where the fat from a roasting animal had fallen into the fire ash, the resulting slimy mess was a good hand cleaner. Wood ash as a paste with water has been used to clean silver and, also, to clean the glass in wood stove doors. In pottery, the discovery that wood ash could be used as a glaze appears to have been a happy accident.

The Chinese potters were the first to make high-temperature downdraught kilns and they noticed that where wood ash had settled upon the rim of cups and bowls during firing, it had formed a dark green glaze.

I have also read of wood ash being scattered in field latrines to help stop the smell, and ashes from a fire – coal or wood – are often used to make an icy path easier to walk on (with wood ash there will be no clinker to clean up later). In Malawi clay pots of beans were once covered with wood ash to keep the weevils out, and I have read that olives can be made palatable through the use of wood ash – the highly alkaline ash neutralizes the bitterness within the olive.

Wood ash folklore

Although there are examples of people rubbing their hair or bodies with wood ash, most folklore seems primarily, and very reasonably, associated with improving the land. There are numerous examples of the ash from spring and midsummer fires being spread onto fields to ensure pest-free crops and to protect crops from the unwelcome attention of witches. As we know, the application of potash is beneficial to a wide range of food crops, so this very ancient practice would seem eminently sensible.

I remember once catching a cockchafer beetle and finding that it was covered in red mites. I put the poor thing in the ashes of an old bonfire and was amazed to see that within a few seconds all the mites had deserted the beetle – which I then fished out again and sent on its way. If I noticed the ash's dust bath/pest-cleaning property, it is very likely that our observant forebears did too. A French custom, which may have its roots in good husbandry, was to take a little wood ash and place it in the nests of chickens.

CHAPTER TEN

CAMPFIRES & WOOD FIRE COOKING

"The glories of a mountain campfire are far greater than may be guessed. One can make a day of any size and regulate the rising and setting of his own sun and the brightness of its shining … Sparks stream off like comets or in round, star-like worlds from a sun."

—JOHN MUIR

For me the legendary Scottish-American naturalist John Muir sums up the joy and delight of a campfire in the words on the previous page. He ignores the prosaic and practical and makes no mention of warmth, drying, protection, or cooking, simply recounting his heart-racing enthusiasm for the campfire's light and sparks.

In this chapter, however, I leave the poetry to John and look at the best ways to make a good, solid, safe campfire, whether in your garden or somewhere in the wilderness. This is not a "survival" book and while I occasionally enjoy the challenge of making a fire from elemental sparks or friction, here we will use matches. Campfires are different from home fires in a few important respects. You can change the shape and size of the campfire to suit the purpose you intend it for. You will, in all probability, be gathering wood for this fire from the trees immediately prior to lighting it. At home you can comfortably manage the fire's draft – now, outside, you have little control over the wind. Don't make your campfire any bigger than it needs to be. This saves fuel and unnecessary work and is the mark of a good woodsman's fire. Lastly, when making a campfire anywhere outside you have a much greater responsibility to control it. An out-of-control house fire will probably only destroy your home, but an out-of-control campfire could destroy many homes.

Let's first look at the fundamental principles of good campfires – keeping it safe, the fuel, lighting the fire, and tending it – and then look at the different basic styles of campfire, and the tools that will enhance your campfire experience. I have enjoyed many excellent meals cooked on a wood fire, so we will also look at how to build a fire for cooking.

KEEPING YOUR CAMPFIRE SAFE

The first consideration must be safety; causing a forest fire is unthinkable. If the area you are in is bone dry, you may be very wise to decide not to light a fire at all. In fact, there is commonly a legal ban on open fires imposed in forested areas during prolonged dry weather. If you are comfortable that it is reasonable to light a fire then choose your fire site thoughtfully and prepare it well. Leaf litter and any surface peat should be scraped away to expose the mineral soil. If the peat or leaf humus is too deep to do this, don't light a fire on it. I have fought forest fires that have gone underground into dry peat and then burned steadily for weeks. If you are by a river, there may be large sandy or stony alluvial flats on the inside of the bends that offer a safe location. In any event, being near water or having some with you is helpful, often essential, when it comes to dousing the fire thoroughly before leaving it – be absolutely certain that it is out.

Preparing a safe fire site

Once you have cleared a patch of ground on which to light your fire, stamp the ground around it to collapse any mice or voles burrows – almost unbelievably the fire can travel along them and escape. We once had a party of Boy Scouts camping in a forest that I managed and they seemed to have done everything just right with their campfire. While they were away on a morning hike a spot fire broke out about 15 yards from their campfire – it was put down to the fire having travelled along one of these tiny burrows. Don't build a campfire against an old tree stump, or a living tree for that matter. The fire may get right into the stump and be very difficult to put out again. Also tree stumps are a great refuge for a wide variety of wildlife that live in dead and decaying wood. Even when you have ensured that the fire site is completely safe, be alert to the nature of the surrounding vegetation. Some plants, such as the grass *Molinia* and gorse, are incredibly inflammable. When camping in fields we were taught to always cut out grass turfs first when preparing the fire site; these turfs were kept to one side to be re-laid later when the fire site was finished with.

Don't be fooled into thinking that putting a ring of rocks around your fire will make it safe – this is a cinema myth. I find more often than not that the rocks get in the way of the fire and many will shatter in the heat. As a young man I used to enjoy swimming with friends in the sea at night, and we would take a straw bale with us to light as soon as we came out from the cold North Sea. As the burning straw blazed, the flint pebbles of the

shingle beach would shatter explosively, with tiny glass-like shards flying at us and burning holes in our clothing. Also, under no circumstances should you resort to using petrol or camping stove fuel. If you really are having trouble getting your campfire going then using a little cooking oil, sump oil, diesel, or thin slithers of rubber with the kindling should help.

WOOD FUEL
FOR CAMPFIRES

The "fuel" is still obviously wood, but differs from the logs you would use at home as it is now probably in the form of sticks and branch wood. For ease, you will most likely collect these sticks and branches from the area around where your fire is to be, so your choice of wood type will be limited to the tree and shrub species growing there. Now your tree recognition becomes important if you are to ensure that you collect the best firewood available. At home, the logs you are burning will have seasoned in your own wood store, or been dried by the merchant you bought them from, but now you must find and identify firewood that has dried naturally, and stayed dry even in wet weather. In all but the driest weather, the golden rule is not to pick up sticks and branch wood from the woodland floor, as these will almost always be damp.

Alternative fuels

It is worth looking out for alternative fuels, such as dung. Once, while camping in a forest in North Wales, I had almost given up trying to light a wood fire as it had rained solidly for days and everything was sodden. Then in a small barn on the edge of a meadow I discovered not dry wood, but the floor of the barn was deep in bone-dry sheep and cattle dung. Not my first choice but on this occasion it made an excellent fire.

Gathering suitable fuel

The trick is to find sticks of a suitable size that are completely dead but still attached to the tree or bush, or sometimes have fallen from a tree and got caught in a bush – anything, as long as the wood is off the ground. If the bark has already fallen off so much the better, but avoid rotten wood – it will smoulder and smoke and ruin a good fire. The more upright a stick, the drier it generally is. If you are faced with gathering campfire wood in the rain, you may find a few dry sticks in the lee of large trees or under evergreen trees with a thick foliage canopy, such as holly or conifers. If it seems hopeless and all the wood is wet, gather smaller sticks than you had intended and make a slightly larger fire. Carry a short length of rope with you while you are out foraging for firewood; it can be used to pull down dead branches that are too high for you to reach and to drag any long pieces of firewood back to your camp. This is easier than cutting up the long pieces where you found them and then carrying the short wood back in armfuls. Another useful tip is to gather more wood than you think you will use, while it's still daylight. It is usually difficult enough to find the wood you need without having to find it by torchlight. Lastly, if rain is possible, cover your wood pile to try and keep it dry.

LIGHTING YOUR CAMPFIRE

We looked at laying a fire ready for lighting in *Chapter 8*, and everything said there still holds true, but now even more care is needed.

It is good practice to always try and light your campfire with one match – it sharpens your focus and skill. Two or three

fig. i

matches would be acceptable, but if the fire has not lit with three matches, then there is probably something wrong with it. If not careful, you may find that you have wasted a great many matches.

Building your campfire

You have made a safe campfire site and have gathered the firewood you need – now think about the wind. If it is very windy you may need to build a rudimentary low windshield to protect the fire and help it to burn usefully. Even if you think there is no wind, just check. On cold days use your breath, on warm days throw a little fine grass into the air and watch how it falls. For me, watching somebody take note of even the lightest breeze is a clear sign that they know what they're doing with a campfire.

If there has been a fire on this site before, then use the fire ash and any residual dry charcoal or burnt ends to form the base for your new fire. If there has not been a fire there before, then make a little platform of sticks, like a tiny raft, as the base for your early fire – this protects it from the damp ground. Remember the latent heat physics we discussed in *Chapter 5*. Now, although it takes heat energy from the kindling's combustion to pre-dry the first sticks of wood, at least the heat is doing something useful. It is not useful to have the kindling's

Campfire

A raft of twigs is created as a base for fire (fig. i); larger twigs are added to create a "V"-shaped wall (fig. ii); small twigs and bark fill the space created ready for the fire to be lit (fig. iii).

fig. ii

fig. iii

precious heat energy drying the ground! Of course the stick platform will eventually burn through and the ground beneath will be dried, but later, when the fire is burning nicely.

I lay two biggish logs on either side of the stick base, parallel to the direction of the wind. I then build a little criss-cross heap of kindling between these logs, taking care to leave a space open to get the lit match in under the kindling. If there is a significant breeze I lie down upwind of the fire and use my body as a temporary windshield while I get the kindling alight. Light your fire on its upwind side so that the first flames are blown into the other sticks.

Kindling options

For kindling I choose a large handful of tiny twigs, about the same diameter as the match that will be used to light them. If there are any birch trees close by, I collect a small handful of the thin bark that is naturally peeling away from the tree's stem – this bark is nature's firelighter. I then tangle this bark among the match-thin kindling sticks. I'm not a fan of grass or leaves as kindling; I want to create a tiny hot centre for my fire and don't find that they help me to do this. An alternative to gathering a mass of match-thin twigs as kindling is to make a few "feather" sticks, or "fuzz" sticks as they are called in Ellsworth Jaeger's excellent book, *Wildwood Wisdom*. Take a dry dead stick and, using a knife or sharp axe, create a mass of little shavings at one end; the shavings must stay attached to the stick to be useful.

It is very satisfying to see your match flame catch in the kindling and the first lively flames grow. Many people now sit back and relax, pleased that their fire is alight – this is a huge mistake. They then look all bemused as it rapidly dies down again and goes out. The point when the kindling is first alight is perhaps the fire's most fragile time and when it needs your closest attention. You must feed each emerging flame a pencil-thin stick and keep a lattice roof of them over the flames, close enough to warm each other but far enough apart to breathe. When these twigs are burning well, you should start feeding the fire with bigger sticks. Only when the bigger sticks are burning can you begin to relax, as now your fire is truly alight.

TENDING
YOUR CAMPFIRE

How you manage your fire and feed it will depend on how dry your firewood is, what style of fire you have made, whether you are mostly after flames or embers, and how you feel about it being smoky. If you are able to split your larger logs do, as split wood burns easier and better. The drier your firewood is, the larger the sticks that can be burned without causing unnecessary smoke, and the fewer you need. Also, a fire burning dry wood is fairly robust as, even if you are distracted and a little late in feeding the fire with fresh wood, the new wood will probably catch easily.

Tending damp wood fires

If you are obliged to burn damp wood, then I suggest you build a slightly bigger fire and put the fresh sticks on lying parallel to each other and the wind. Damp wood fires will often burn out a hollow dome within the fire. Watch out for this and don't let it happen, as the fire will begin to die once the hollow has formed. Tending a fire in a wet environment is a constant challenge, while tending a fire in a very dry environment is delightfully simple.

I have only rarely wanted my fire to be smoky, and this was when clouds of biting midges or mosquitoes were plaguing me. A friend of mine in Ireland would concoct a special anti-midge smoking mixture for his campfire, which was made up of dried horse dung and a lovely little plant found in boggy places called sweet gale or bog myrtle (*Myrica gale*). I have sat in the smoke cloud he created on still, drizzly days when the midges were fierce, and it did seem to work.

STYLES OF
CAMPFIRE

There are four basic campfire designs: pyramid, star, square, and long fires. There are also curiosities such as the post-top fire and a fire reflector adaptation to throw the fire's heat back at you. Once you have a basic fire alight and burning well, you can begin to adapt it to the design that best suits your purpose. As I have said earlier in this book, a rough heap of burning wood is nothing to be proud of, you can do better than that – get organized and build a fire that really does what you need it to.

The pyramid fire

In this design the fire is fed to form a round cone. Build a low cone, as steep ones will tend to keep falling over. These fires are good for flames and light, just what you need to sit around with friends of an evening. They are not good for general cooking, but are perfect for boiling water. How many films have I seen with Robin Hood or some cowboys spit-roasting a small animal in the flames of a little pyramid fire – this is absolute rubbish. The outside meat would blacken and burn, while the inside would remain raw. Boiling is a different matter; a pot or kettle suspended in the top of, or just above, the fire's flames is fine and should boil quickly, especially if you feed the fire a steady diet of small, dry sticks. When I was 16 and first started working in woodlands, I was given the job of getting the tree-felling gang's campfire going for their lunch break – they wanted a boiling kettle and the chance to toast their sandwiches. Luckily I had had a boyhood of playing with matches and could light a reasonable pyramid fire quickly, but this was still a tough proving ground for me – the men were rough old woodsmen and woe betide me if my fire was feeble when they arrived for their break.

The star fire

This is a lovely, simple design and is widely used when the firewood is reasonably dry. Large-diameter long logs are placed to form a star radiating from the fire. The ends of each log will begin to burn, forming a very compact fire. To start with, and when it needs perking up a little, feed smaller sticks into the centre of the star fire. If your large logs are damp this fire will never burn well and will smoke a lot. The logs are gently pushed into the fire as their ends burn away, which saves a lot of time cross-cutting the firewood into short lengths. The star is sometimes called a "lazy" fire, and is a good design for cooking as a pot can be suspended above the fire, or a metal grill rested across the star logs as a good base for frying pans, kettles, and pots. If it is cold, however, this is not a great design for sitting around with friends. The star logs tend to shield the embers from you and get in the way, with people tripping over or knocking them. Also, the star design does not produce much flame.

The post-top fire

I first saw these post-top fires used in Finnish Lapland at a cross-country skiing event in deep snow when the air was just below -20°C (-4°F). Dry posts 20–30cm (8–12in) in diameter had been driven into the snow so that the post top was about waist high. The post tops had been cut vertically (rip sawn) down to about 23cm (9in) across the centre three times. This divided the remaining wood into six pie-shaped segments that were still all attached to the post at one end. An oily rag was pushed into the centre of the six segments and the fire lit. These small fires burned well for what seemed a very long time and were at a perfect height for hand warming.

Something else I saw that day that impressed me was a fire pit full of children. It is difficult for really small children, with their small body masses, to handle severely cold weather and so the

Post-top fire

At the correct height for hand-warming, the post-top fire is effective and quick to light with the aid of an oily rag.

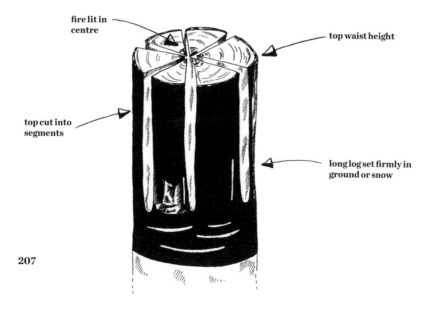

fire lit in centre

top waist height

top cut into segments

long log set firmly in ground or snow

toddlers' parents had cut them a crèche in the snow, roughly 1m (3ft) deep. They lined one half of the pit with reindeer skins and put an oil drum full of burning pine logs on the other half. I watched the little children playing very happily together on their reindeer skin bed – they seemed well aware that they needed to keep away from the hot, slowly sinking oil drum.

The square fire

This is a useful all-round design as it produces a pleasing balance of good embers and flames. The well-made square fire should throw out a lot of low heat and is good to sit around. I've seen this design also called a "council" or "criss-cross" fire. Once the initial fire is going, larger logs are built around it, forming a low trapezium pyramid. You should quickly end up with a good bed of embers within the square, with the outer logs burning steadily. I always place the first two largest logs parallel to the wind direction so that air can move through my fire freely. As with the star fire, this is a good design for cooking as there are plenty of embers.

My honeymoon was a gold prospecting trip in northern Scandinavia. While hiking along a large river in the sparse boreal forest we came across a recent Sámi camp. Two things impressed me: the first was that they had put a huge effort into making a thick, springy bed mattress of tiny juniper twigs, and the second was their cooking fire. They had made a standard square fire, but with a forked stick pushed into the ground on each corner. Two thick, straight, freshly cut poles were laid between each of the pairs of forks. And then the clever thing – two slightly thinner poles were loosely laid across the primary two. These gave support for a pan or grill and the cook had the complete flexibility to move the cooking food to whichever part of the fire was the right temperature.

The long fire

There will be times when you want to cook a large meal and there simply is not enough room on a normal round or square fire to take all the pots: in this instance, a long fire is needed. I first enjoyed a long fire on the island of Hokkaido in northern Japan. I had been invited to eat

with a Japanese family and when I arrived I found that they had made a long fire in a shallow pit, about 4m (13ft) long, so that all the guests could sit comfortably and keep warm. I used the idea when I took my children on a desert trip in Morocco; we were on the northern edge of the Sahara desert over Christmas and I wanted us to have a proper Christmas dinner. We took with us turkey, potatoes, vegetables, and gravy, as well as Christmas pudding and custard, and built a beautiful long fire in order to cook and enjoy it all. It was in Morocco that I first noticed how well a fire burns in the desert. The scarce wood is very dry and of course it burns easily, but it was the embers that most impressed me. They just glow and glow and last for ages – I assume this is because the ground beneath the fire is so dry.

A variation of the long fire that is designed for cooking involves having flaming sticks at one end of the fire producing fresh embers for the other end, which is where the cooking takes place. The cooking end of the fire consists of just ashes and embers, with new embers being added as required to keep the heat up. There is another situation in which you may need a long campfire, and this is for sleeping or surviving in extreme cold. A friend of mine in Arctic Lapland, who hunts ptarmigan in winter and fishes for salmon and gold dust in summer, once showed me how to do this. Two very dry logs around 1.8m (6ft) long are selected and laid close together – about 8–13cm (3–5in) apart – on a bed of smaller wood. A fire is then lit in the gap between them along their entire length. In the far north the best wood for this was ancient dead pine – cut from trees that were still standing with all their bark long gone, leaving the hard, resinous bare wood weather-bleached and grey.

Fire reflectors

If you are only sitting on one side of the fire, then it may be an idea to build a reflector on the other side. This is easily done by pushing two sharpened poles into the ground, which should lean slightly away from you. Place split logs horizontally against these uprights, with the clean, white split surface facing you. These reflectors significantly increase the warmth you will feel from your fire.

CAMPFIRE
TOOLS

Storm kettle

An ingenious and ecologically sound invention providing heat and hot water from very little fuel, designed especially for use in bad weather conditions. The cork must be out when boiling water.

What tools you choose to carry with you will to a large degree depend on the nature of your campfire. I generally carry a small hand axe and bow-saw. On camping trips I used to carry a very light saw blade and then make the bow from a bent stick once I was on-site. Now I tend to take a folding pruning saw with me, as this will cut everything I want and is safer and easier to use. If weight is not a problem, then having a couple of iron bars or a metal grill to place over the fire as a support for cooking pots is a huge help.

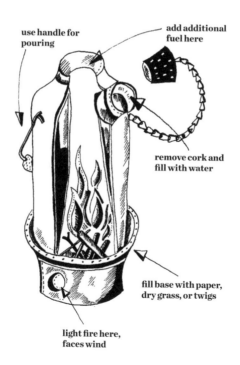

use handle for pouring

add additional fuel here

remove cork and fill with water

fill base with paper, dry grass, or twigs

light fire here, faces wind

211

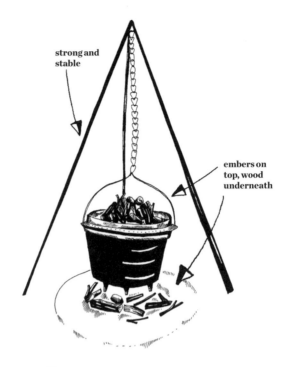

Dutch oven
A thick-walled, usually cast-iron pot with tight-fitting lid, the Dutch oven has a concave lid, allowing embers to be placed on top of the casserole for a more even heat distribution.

strong and stable

embers on top, wood underneath

Campfire cooking utensils

There is not such a wide range of special campfire-side equipment available as there is for home fires. Three things that could be really useful are a wood-fired storm kettle, a Dutch oven, and a grill that folds over to lock fish or steaks in place. The storm kettle is a lovely little thing, with the water container sitting over an enclosed fire space. The tiny fire is fed with twigs, pine cones – even grass and leaves – and the water does seem to boil quickly. Be sure to read the manufacturer's advice as some kettle designs come with a whistle, and some have a cork which must be removed before boiling. The cast-iron Dutch oven is a wonderful piece of cookware that slowly cooks gorgeous stews, or can be used dry for baking, while locking grills make cooking over embers very easy.

If you like outdoor cooking you will soon learn that every form of cooking can be done with the campfire – boiling, grilling, frying, stewing, baking, and roasting. Sometimes it is fun to improvize and I have baked gorgeous pizzas in two metal gold pans, which are a bit like a wok or frying pan, with one pan placed over the other to form an oven and set a few inches above the white ash of a mature fire. You can also wrap food in aluminium foil to cook in the ashes of an old fire, or make a fire pit to bake your food slowly. To encourage camp hygiene, I have also lit a twig fire in a rough chimney made of riverside rocks and placed a small metal bowl of water over it, so that my children could wash in warm water – perhaps I spoiled them!

TO FINISH

Thank you for reading this book. I hope that there was something useful to you within its pages. While writing, I have journeyed back through my past, triggering so many thoughts and happy memories of times and people. I have perhaps been spoiled by the privilege of having had the companionship of wood fires all of my life. I've likened my fires to having a much-loved pet, or perhaps a scrap of barely tamed wildness in my home. I feel sympathy with the ancient Greeks, who so neatly classified earth, water, and air, but struggled with how best to define fire. I hope that I have touched on all the many ways in which wood fires can enrich our lives, and that we, who warm ourselves with this most ancient fuel, are still modern and responsible. I am, as I sit here writing this very last line, reminded of words that J. R. R. Tolkien wrote for Bilbo Baggins:

"I sit beside the fire and think of people long ago,
and people who will see a world that I shall never know"

– As do I.

*

APPENDICES

The gain in available heat energy as logs 'season' and dry out

Logs are considered ready to burn at 20% moisture content. I tested a full sample of the naturally seasoned logs I was burning in late February 2019, and they averaged 16% moisture content. I live in the east of England, with a fairly dry climate. By properly drying my firewood, I have gained more than double the heat energy from the logs, and reduced the emissions from my stove about six fold – and not caused damage to my chimney or stove.

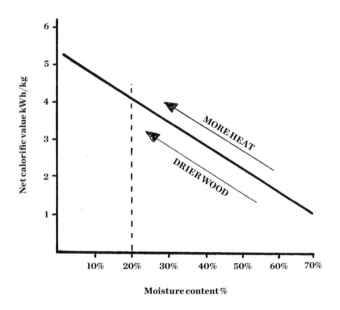

Source: Forest Research, Forestry Commission

Indicative costs of main heating fuels

The price for oil and wood seems the most fluid. Three years ago, heating oil was half the price used below. Firewood prices are very variable and difficult to quantify – but it is the only fuel where a home owner commonly collects some as a contribution to their own home's heating need. For firewood I have used a price of £250/t for logs at 20% moisture content. 'My wood' allows for my collecting 25% of my own firewood and buying in 75%. Also, my firewood is slightly drier at an average of 16% mc. I have not shown house coal as further restrictions on burning this cheap and polluting fuel are likely. Of the heating fuels shown, locally sourced and sustainably produced firewood is almost carbon neutral. Gas, oil, and coal are obviously fossil fuels releasing ancient carbon if burned. Electricity production, in the UK in 2018, was roughly 33% renewables, 20% nuclear, and 47% fossil fuels, primarily gas. This mix is why electricity is often described as a 'low carbon' fuel.

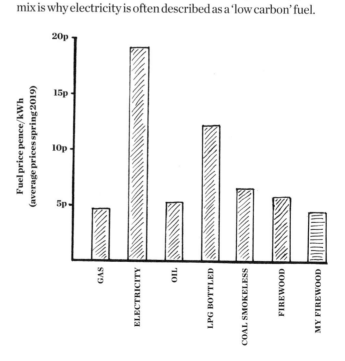

INDEX

Page numbers in **bold** refer to main references (including their photographs and illustrations), *italics* refer to all other photographs, illustrations, and captions

Æsce 193 (*see also* wood ash)
air-seasoning 84
air-venting 129, **157**, 158
alder (*Alnus*) **49**, 71, 83, 85, 100, 105, 164
andirons 31
apple (*Malus*) 44, 51, 100, 110, 146, 166
archaeology 11, 63, 65
ash (*Fraxinus*) 42, **49**, 71, 75, 96, 100, 105, 110, 124, 134, 164, 166, 173, 182
aspen (*Populus*) **54**
axes 21, 34, 37, 50, 53, 62, 63, 72, 122, 132–8, **139–40**, 141–3, **144–5**, *145*, 146–9, 151, 178, 204, 211

back boilers 26
backdraft 187
bark 41–42, *41, 44, 45*
bark-free wood 98
basswood (*Tilia*) **55**
bay (*Laurus*) **47**
bed-warming pans 192
beech (*Fagus/Nothofagus*) 24, 25, 42, **43–4**, *45*, 52, 71, 83, 100, 105, 110, 124, 133, 134, 164, 166, 182, 190
beehive stacks 125, 126
bellows *30, 31, 34, 161, 173*
Ben (woodsman) 99
benders 32
billhooks 72
bins, metal 32, 91
birch (*Betula*) 16, 21, *44, 45*, **50**, 71, 79–80, 99, 100, 105, 110, 134, 146, 161, 164, 168, *168*, 173, 182, 204
birch-bark links 168

bird lime 52
black locust (*Robinia*) *44, 45*, **46**, 100, 164
blackthorn 100
blow pokers 32, 173
bog oak (*Quercus*) 124
bonfires 11, 37, 46, 179, 187, 188, 192, 195
"bottom-up" approach 170
bow-saws 72, 211
box (*Buxus*) **47**, 100, 166
box elder (*Acer*) 53
briquettes 18, 79, 91
broadleaves 35, 37, 42, 55–7, 69–71 (*see also by name*)
brushes 31
butane lighters 172
butternut (*Juglans*) **53**
buying firewood 20, **76–93**

campfires 9, 21, 187, 189, **196–213**
carbon cycle 35, 36
carbon dioxide 180, 187, 189
carbon footprint 36
carbon monoxide 164, 180, 189
carbon neutrality 36, 96, 217
cedar (*Cedrus*) 55, **57**, **58**, 102, 113
chainsaws 21, 37, 44, 62, 72, 74–5, 126, 138, 143, 146, 151
charts **216–7**
cherry (*Prunus*) 51, 100, 166
chestnut roasters 34
chimney(s) 15, 16, 19–20, **24–5**, 26–31, 57, 97, 126, 154, 155, 160, 170, 183, 187, 216
chimney bricks 25, 175
chimney-damaging acidic tar/creosote 158, 159
chimney design 163
chimney draft 164–5
chimney emissions 186

chimney fires 159
chimney flue 157, 165
chimney lining 100–2, 159, 186
chimney sweeping 19, 87, *161*
chopping blocks *117*, 132, 134, 135, **142–3**, 144–9 *passim, 145*
A Christmas Carol (Dickens) 158
Christmas trees 9, 58, 224
Cider with Rosie (Lee) 13, 155
clear-felling/clear-cutting 68, **69**, 70, 71
clinker 15, 163, 193, 195
clinker-free ash 193
coal 15, 24–5, 28, 32, 36, 66, 77, 78, 81, 84, 86, 110, 158, 163, 166–7, 187, 193–5, 217, *217*
coal ash 15, 195
coal in fiction 110, 158
coal grates 163
coal mines 110
Cold Comfort Farm (Gibbons) 110
companion sets 31, *161*
conifers 35, 42, 50, **55–9**, *56*, 63, 69, 70, 71, 74, 85, 100–3, 113, 134, 148, 164, 202
cooking 10, 13, 32, 34, **160–1**, **167**, 173, 190, **196–213**
cooking utensils **212–13**
coppicing 51, 66, 68, **70–1**, 72, *73*, 75, 97
copses 36
The Cottage Homes of England (Dick) 28
cottonwood (*Populus*) **54**
creosote 91, 158, 159, 186, 188
crotch 134, 145, 146, 149
cypress family (*Calocedrus, Chamaecyparis, Cupressocyparis, Cupressus*) 55, **58**, 102, 113

Del Mar, Alex 29
Dick, Stewart 28
Dickens, Charles 158
diffuse porosity 100,
133, 134
"dirty" wood 91
dogwood (*Cornus*) 42,
47, 80
domestic heating oil 15,
120, 217
Dryden, John 153
Dutch elm disease 190
Dutch ovens 160, 212, *212*

electric 15, 18, 32, 78, 86,
92, 106, 123, 129, 151,
166, 179
elm (*Ulmus*) 25, **44**, 100,
105, 158, 164, 190
embers 9, 34, 43, 44, 46,
80, 157–9, 163, 165–7,
172–3, 175, **176–95**
emissions, regulation of
14, 19
engineered wood fuels
91–2
eucalyptus (*Eucalyptus*)
35, **50**, 52
euphorbia (*Euphorbia*) 42
Evans, Sir Arthur 65
evergreens 42, 202 (*see
also* conifers; *see also
by name*)

faggots 173
felling trees *see* tree-
felling
fenders *30*, 31, *31*
field maple (*Acer*) **46**, 164
fir (*Abies*) 55, **59**
fire **152–75**, *181*, *184*
fire backs *30*, *31*, 32, *161*
fire bed 163
The Fire of Drift-wood
(Longfellow) 177
fire extinguishers 34
fire-feeding 14, 190, 204,
205
fire, first tamed 10–11
fire reflectors 205, **210**
fire screens 34
fire-tending 14, 80, 165,
198, **205**
firedogs *30*, *31*, *161*, 163
firelighters 155, **168–73**
fireplace dampers 165

fireplaces 16, **25–7**, 32, 34,
80, *161*, 164, 165, 170, 193
fires in fiction 13, 110, 155,
158, 166
fireside accessories **31–4**,
30, *31*, *34*
firewood 15, 18, 23, 25, 31,
35, 37, 40–2, **43–59**, 66,
68–9, *70*, 71, 75, **76–93**,
94–107, **108–29**,
130–51, 154, 156, *156*,
159, 164, 186–90, *191*,
196–213, 216, 217
firewood bought locally
20–21, 36
firewood harvesting 35–7,
64, 66, **68–72**, *73*, 74,
126, 188
flames 153, 154, **176–95**,
196, *200*, *209*
flue thermometers *156*
folklore 47, 173, **188**, **192**,
195
fossil-fuel footprint 36
fossil-fuel reserves 96
fossil fuels 11, 36, 37, 96,
217 (*see also* coal; gas;
oil-fired heating)
"free" wood 21
frozen wood 134–5, 140,
145
fruitwoods **51**, 80, 100,
190 (*see also by type*)
"fuel and furze" day 21–3
fumigation 29, 188
furze (gorse) (*Ulex*) 21,
29, 199

garden trees 57, 68
gas 15, 36, 78, 44, 107,
217, *217*
Gibbons, Stella 110
gloves *30*, *31*, 32, 141
gorse (furze) (*Ulex*) 21,
29, 199
grass 199, 203, *211*, 214
"green" wood 18, 43, 44,
46, 49, 51, 79, 81, 84, 88,
96, *103*, 106, 107, 119,
133, 134, 145, 186
growth rings *41*, 42
Gymnosperms 55

half-burned logs 165
hard maple (*Acer*) **46**, 166
hard softwoods 42

hardwoods *41*, 42, 57, 83,
85, 88, 97, 100, 105, 193
(*see also by type*)
Harrison, William 29
hawthorn (*Crataegus*) **46**,
100, 166, 182, 190
blackthorn (*Prunus*) 100
hazel (*Corylus*) **51**, 71, 75,
100, 164, 173
hearth 16, 19, **24–9**, 31, 34,
91, 110, 163, 167, 189 (*see
also* fire; open hearth)
heartwood *41*, 43, 44, 46,
49–55 *passim*, 106
heat transference 175
hedgerows 36, 46, 68, 190
hemlock (*Tsuga*) 55, **59**,
102
Henry VIII 71
hickory (*Carya*) **47**, 100
Hill, Archie 170
holly (*Ilex*) **51–52**, 100,
202
holm oak (*Quercus*) **47**,
100
Holz Hausen open stack
99, 126
home, suitability for wood
burning 15, 18–19
hornbeam (*Carpinus*) 25,
44, *44*, *45*, 71, 80, 100,
164, 190
horse chestnut (*Aesculus*)
42, 52, **54**, 80
house coal 217
Huth, Angela 46

identification **40–2**, *44*,
45, 50, 87, 129, 201
"intelligent laziness" 134

Jaeger, Ellsworth 204
juniper 59, 208

kettles 21, 32, *156*, 160,
167, 186, 206–7, 211,
211, 212
kiln-drying 79, 84, 96,
110, **129**, 187, 188, 195
kindling 24, 34, 43, 54, 57,
83, 87, **90–1**, **148–9**, *156*,
161, 164, 167, 170–3, *171*,
201–4, *202–3*
kitchen-roll inner-tubes
91

laburnum (*Laburnum*) 42,
  47, 100
The Land Girls (Huth) 46
landfill 11, 37
larch (*Larix*) 55, **58**, 79,
  102
latent heat 97, **107**, 203
laurel (*Laurus*) 42, 95
Lawson cypress
  (*Chamaecyparis*) 113
"lazy" fire **206–7**
leaves 14, 47, 98, 119, 153,
  172, 189, 204, 212
Lee, Laurie 13, 155
Leyland cypress
  (*Cupressus*) 113
lighting, wood store 123
lime (*Tilia*) **55**, 100, 194
linden (*Tilia*) 55
local economy 16, 18,
  25, 36
log baskets/holders/
  carriers 16, *30*, 31, *31*
log fire *2*, *6*, 10, *12*, 16, *29*,
  30, 62, *152*, *162*, *174*, *181*,
  191, *214*
log identification **41–2**
log seasoning *see* seasoned
  wood
log shapes **145–8**
log-splitters, powered **151**
log-splitting *see* splitting
log-stacking **83–5**, 86, 87,
  94, 99, *101*, 119, **121–2**,
  125–6, *127*, *148*
log wall 121–2, *121*, *122*, 133
logs 16, 94, **94–107**,
  130–51
logs, difficult to split 46,
  27, 51, 134, 137, 138–9,
  142, **145–8**, *146*
logs, ready-cut 24
long fire 205, **208–10**
long log storage **126–8**
longfire 214, 216
Longfellow, Henry
  Wadsworth 177

maple syrup 98
maples 42, **46**, **52–3**, 71,
  100, 164, 182
match holders 32
matches 32, 167, 170, 172,
  198, 203, 206
medullary rays *41*, 42, 43,
  44, 46, 52, 53

mesquite (*Prosopis*) **47**
microchip-controlled
  sensors 91–2
moisture meter 105,
  106–7
Molinia (grass) 199
Muir, John 197
*My Wood Fires and Their
  Story* (Robinson) 26, 166

Newcastle coal 28
newspaper *see* paper
Norway maple (*Acer*) 98
nutshells 91

oak (*Quercus*) 25, 42, **43**,
  44, *45*, **47**, 52, 64, 71, 78,
  83, 98, 100, 105, 106, 110,
  120, 142, 146, 164, 166,
  170, 190
off-cuts **90**
oil-fired heating 15, 120,
  217
one-match fires 21, 170,
  202
open hearth 9, 79–80, 132,
  152–75 (*see also* fire;
  hearth)
orange peel 91, 168, *168*,
  173
overnight burning **159–
  60**, **167**

paper 69, 81, 157, 168–70,
  169, 211
paper spills 32
paper sticks 169
parks 52, 57, 68
pear (*Pyrus*) 51, 100
peat fires 183–4, 199
pellets 18, 28, 40, 79, 91–2
perfect fire 80, 91–2,
  161, **152–75**, **176–95**,
  196–213
petrol 151, 201
The Picture of Dorian Gray
  (Wilde) 110
Pine (*Pinus*) 35, *44*, *45*, **58**,
  79, 80, 102, 105, 148, 161,
  182, 208, 210
pine cones 91, 168, 212
pizza ovens 44, 212
plane (*Platanus*) **52**, 100
pokers 31–2, 165, 173
pollarding 65, 68, **72**
polytunnel woodshed 129

poplar (*Populus*) 35, 42,
  54, 55, 79, 80, 85, 100
porosity 100, 133, 134
post-top fire 205, **207–8**
potash 193–5 *passim*
powered wood-splitters
  151
Practical Forestry
  (Webster) 138–9
PTO 151
Punch 77
purification 63, 188
pyramid fire 212

question, for your
  firewood supplier **87–8**

raised fire bed/grate/
  hearth 25, 29, 163, 167
ready-cut logs 24
"ready-for-burning"
  target 96
red cedar (*Thuja*) **58**, 113
red oak (*Quercus*) 43, 100
redwood (*Sequoia*,
  Sequoiadendron) 55, **59**,
  70, 102
register plates 26–8, 29
Rendlesham Forest 20,
  113
rhododendron 42
ring porosity 100
Robinson, W. 26, 28, 166
rock chimney 213
rock feldspar 169
rock maple (*Acer*) 46
rocks 199, 125, 213
rowan (*Sorbus*) **47**, 100
Royal Forestry Society
  (RFS) 90, 137

sapwood *41*, 42–55
  passim, 106
sawdust and shavings
  14, 92
sawmills 20
saws 34, 72, 75, 81, 134, 211
  (*see also* chainsaws)
Scots pine (*Pinus*) *44*, *45*,
  102, 161
Scottish oak (*Quercus*) 64
screening 63, 124
sea coal 25
seasoned wood 14, 18, 19,
  44, 49, 53, 58, 59, 79, 81,
  84, 88, 96, **97–103**, 106,

107, 113, 119, 126, 134, 156, *161*, 178, 201, 216
seasoning time 59, **97–103**
The Second Meadow (Hill) 170
secondary burn 155, 170, 186, 189
shavings and sawdust 14, 92
solar kilns 110, **129**
shovels 31
silver birch (*Betula*) *44, 45*
slabwood **90**
sledgehammer 138, 141, **149–51**
small transportation footprint 36
smoke **176–95**
smoke scent 15, **42**, 43–59 *passim*, 102, 110, 146, 161, 183–5
smoke signals **187–8**
smokeless fuel (coal) *217*
smokeless zones 19
social capital 23–4
soft maples (*Acer*) **52–3**
soft softwoods 42
softwoods 42, 57, 88, 90, 97, 193
soil regulation 11, 63
solar kilns 110, **129**
soot 19, 29, 50, 57, 58, 159, 180, 186
spark guards 34, *161*
spill holders 32
spinneys 36
splitting 46, 27, 51, **130–51**, 154
spruce (*Picea*) 35, 55, **58**, 59, 80, 102
square fire **208**
star fire **206–7**, 208
stacking **83–5**, 86, 87, *94*, 99, *101*, 119, **121–2**, 125–6, *127, 148*
storage *see* wood storage
storm kettles *211*, 212
"The Story of Baucis and Philemon" (Dryden) 153, 154
stove thermometers 32
stovepipes 28
stovetop fans *30, 31*, 32
stovetop steamers 34
street trees 36, 52, 68

sugar maple (*Acer*) 46
sumac (*Rhus*) 42
sweet chestnut (*Castanea*) 34, **52, 54**, 71, 79, 100
sycamore (*Acer*) **52–3**, 100, 164

tar 158, 159, 186, 188
tarpaulins 117, **128–9**
thermocline 32
thermometers 32, *156*
thinning 68, **69**, 125
tinder 155, **168–70**, *171*
toasting forks 34
toilet-roll inner-tubes 91
tongs 31, *34*
"top-down" technique 170–1, *171*
tree identification **40–1**
tree-felling 35, 50, 62, 65, 68, **69**, 70–1, **72–5**, *73*, 98, 102, 139, 140, *140*, 144, 206
tree-planting 35, 65, 66, 70–1
trees **38–59** (*see also* trees by name)
trivets 32–3, *156*, 160
trugs 31
tulip tree (*Liriodendron*) *55*
"turf" fires 183–4, 199
Twain, Mark 166

under-floor draft 25, 26, 157, *161*, 164
unseasoned wood *see* "green" wood
US Forest Products Laboratory 99

walnut (*Juglans*) **53**, 100, 166
warming pans 192
Warner, Charles Dudley 166
water regulation 11
Webster, A.D. 138–9
wedges **149–51**
Western red cedar (*Thuja*) **58**, 113
what is a flame? **179–82**
what is wood smoke? **185–7**
white ash 14, 167, 193, 213
Wilde, Oscar 110

wildlife 36, 63, 69, 114, **124–5**, 138, 199
Wildwood Wisdom (Jaeger) 103
wildwoods 5, 63, 74, 204
willow (*Salix*) 35, 42, **53–4**, 71, 79, 80, 83, 85, 100, 105
winter firewood 66 (*see also* firewood)
wood ash *161, 171*, 186, 189, **193–5**
wood boilers 28, 92
wood briquettes 18, 79, 81
wood-burning stoves 14, 16
wood fires lifestyle **12–37**
wood pellets 18, 28, 40, 91–2
wood scent 87 (*see also* smoke scent)
wood storage 16, 18, 28, **31–2**, 57, 79, **83–5**, 100, **108–29**, 137, 201
wood tar 158, 159, 186, 188
woodchips 18, 79, 92
woodcutting **60–75**
woodland(s) 9, 11, 14, 21, 35–7, 40–1, 46, 62–75, 87, 98, 113, 114, 125, 135, 138–9, 141, 143, 170, 180, 201, 206
woodland associations 87
woodland charities 90
woodland walks 90, 91
woodsheds 31, 44, 50, 59, 86, 96, 99, 105, 106, 110–13, **114–24**, 129, 133, 148, 157
wood smoke **42–3**, 146, **183–90**
"Woods to Burn" (Anon.) 51
Wright, Frank Lloyd 39
wych elm (*Ulmus*) 44

xylem vessels 44

yellow poplar (*Liriodendron*) **55**
yew (*Taxus*) 42, **47**, 100
yule logs 161

Picture credits

Acknowledgements

A lifetime with wood fires has brought me into contact with many wonderful people, but I would first thank my family. I have dedicated the book to my parents but would also thank my children Daisy, Jack, and Ralph and in particular my partner Mary for her tireless support and patience as I've tapped away writing this, my first book, for countless hours at home and in tents and cabins while we were away on gold prospecting trips.

I am also truly grateful to Jane, my wonderful agent, who has so carefully guided my first steps into the world of publishing and who immediately understood my love of fires.

I am particularly indebted to all those at Octopus Publishing Group who have so brilliantly turned my rough manuscript into such a beautiful book, especially Denise, Jo, and Pollyanna for their support, patience, and enthusiasm which helped me hugely. I would also like to thank my daughter Daisy and Meg for helping with the research and Darren for his quiet professionalism in producing the videos that support the book.

For many years I have tried to turn every evening conversation, whether at dinner, in my village pub or by campfires, to the subject of wood fires. So many people have indulged me and been happy to share their knowledge with me. I would like to thank in particular: Ben and Sylvia Campbell, Tony and Roz Penrose, Robert Rippengal, Adam Green, Charlie Field, Alf Henderson, John Shipp, Diana and Richard MacMullen, Gypsy Jim's grandson Jim, James Ogilvie, Tim Curtis, Greta and Charlie Clark, Mike and Colleen Jones, Trevor and Miriam Mason, Damian and Justine Conway, Andrew Falcon, Chris Halls, and Rob Chapman.

And finally, the men and women of the Huntley Estate and Dartington Hall who gave this young man with his passion for woodlands and fire his first steps into the woodsman's world.

Vincent Thurkettle is a woodsman.
After spending his childhood roaming
the Welsh and English countryside, he
left school at 16 to work in woodlands.
He subsequently trained as a Chartered
Forester, and worked for the Forestry
Commission. Vincent now runs his own
Christmas tree business and still heats
his home exclusively with the wood he
cuts each winter.